T0291516

Modeling and Benchmarking Supply Chain Leadership

Setting the Conditions for Excellence

Series on Resource Management

Modeling and Benchmarking Supply Chain Leadership

Setting the Conditions for Excellence

Joseph L. Walden

CRC Press
Taylor & Francis Group
Boca Raton   London   New York

CRC Press is an imprint of the
Taylor & Francis Group, an **Informa** business

Auerbach Publications
Taylor & Francis Group
6000 Broken Sound Parkway NW, Suite 300
Boca Raton, FL 33487-2742

International Standard Book Number: 978-1-4200-8397-2 (Hardback)

Library of Congress Cataloging-in-Publication Data

Walden, Joseph L.
    Modeling and benchmarking supply chain leadership : setting the conditions for excellence / Joseph L. Walden.
        p. cm. -- (Series on resource management)
    Includes bibliographical references and index.
    ISBN 978-1-4200-8397-2 (alk. paper)
    1. Business logistics--Management. 2. Benchmarking (Management) 3. Leadership. I. Title.

HD38.5.W34 2009
658.5'03--dc22   2009017401

Visit the Taylor & Francis Web site at
http://www.taylorandfrancis.com

and the Auerbach Web site at
http://www.auerbach-publications.com

Contents

SECTION II: LEADERSHIP©: THE ATTRIBUTES OF LEADERSHIP

SECTION III: APPLICATIONS

Acknowledgments

I would like to thank every person for whom I have worked. From each of these people I have learned about leadership — some of them showed me how to be a caring, serving leader and some of them showed me what not to do as a leader. Larry Matthews, Al Stein, Joe Gaglia, General Claude Christenson, and Lieutenant General J.D. Thurman not only taught me how to be a leader, but also set the example for how a leader should act. Special thanks to a couple of other bosses, who will remain nameless, for their examples of how not to lead effectively; I may have learned more from them than from the ones who taught me how to lead.

My greatest example of how to be a leader is my father, Thomas Walden, Sr. I watched carefully as a youngster as he served as Scout Master for the Tennessee School for the Blind. He was a man ahead of his time as he integrated the Boy Scout troop and several times a year the Shiloh Methodist Church. He also taught me the importance of giving of oneself for others as the Scout Master and as the volunteer pharmacist for the McKendree Manor Nursing Home. Most importantly, he also showed me that no matter how busy you are, there is always time for your family and friends. Perhaps one of the greatest lessons he still teaches every day is that you can talk all you want about values, principles, and leadership but the best way is to demonstrate your beliefs is in your daily life.

My lessons in leadership continue every day as a father of the two most wonderful daughters any father could ask for — Amber and Bobbi.

Without the constant support of my wife Kay, I would probably be living the life of a beach bum in Hawaii — which was my plan before we decided to marry.

Another of the greatest lessons in life came from Don Ho. As a young officer, Uncle Don told me that you are never too old to kiss the people that you love or too old to tell them, "I love you." Too often as we grow older, we start thinking that the important people in our lives know that we love them and do not need to hear it. Thanks for the memories, the lessons, the Aloha, and the fun. Don and his wife Haumea always modeled the lessons of Aloha. Uncle Don Ho passed away as I began to write this book; he may be gone but his lessons continue.

I have been blessed with a great family that has always believed in what I wanted to do and have been lucky enough to have shared time with, learned from, and broken bread with two American legends — Uncle Don Ho and Buck O'Neil.

O'Neil was a true gentleman, always willing to share his experiences and lessons and always ready to give a hug to every lady in the house — with a smile and a "give it up," he quickly lit up any room into which he walked. O'Neil had the ability to get grown men to hold hands and sing along with "ol' Buck" the words to his favorite song. One of my greatest accomplishments as a brigade commander was to introduce O'Neil and his stories to the soldiers at Fort Irwin. I had the honor of hosting him in my house for breakfast before his talk to the soldiers of the Theater Support Command. His first visit was so popular that while I was vacationing in Kuwait, the post Equal Opportunity Office brought him back again to share his stories and inspiration with all the soldiers of the National Training Center.

As with my first book, all proceeds from this book will be donated to the Negro Leagues Baseball Museum in honor of Buck O'Neil, a true American legend and pioneer. If you are ever in Kansas City, Missouri, please take the time to visit the Negro Leagues Baseball Museum and spend a few hours absorbing the history of America.

Special thanks to my close friends Barry Walker, Melinda Woodhurst, and "Uncle Billy" Pratt for providing me with a sounding board for ideas and letting me know when I was off track. They will never know how much their friendship over the years has helped me keep going.

Mom: I love you and miss you every day.

Foreword

Supply chains are inherently complex, extended, and dynamic. Because of these factors found in every supply chain, leadership is required to guide the people who make up the supply chain organizations through the problems created by global, complex, dynamic supply chains. The military theorist Karl von Clausewitz wrote about the fog and friction of war. In supply chains there is the same fog and friction caused by the inherent nature of today's supply chains and the necessity of meeting the needs of a constantly changing customer. Leadership in the supply chain industry requires supply chain leaders. In this book we look at the individual leaders needed to make supply chain companies leaders in their fields. In doing so, we introduce a new problem facing leaders in all areas of business, not just supply chains. This problem is *motivational dysfunction*. The cause of motivational dysfunction is lack of leadership. The results of motivational dysfunction can be seen in products and offices throughout the world — employees surfing the Web, chatting over the Internet, not producing at the required levels of performance, talking on the phone about personal business, and basically costing companies millions of dollars a year in lost productivity. I recently visited an organization that had placed more than twenty television channels on its intranet and made them available to every employee. How many do you think were wasting the day watching television over the intranet versus doing what they were being paid for?

The cure for motivational dysfunction is leadership. We are all in the people business and when dealing with people, leadership is required. Management is for assets such as inventory and capital. People need leadership; and with proper leadership as discussed in this book, motivational dysfunction can be cured in every organization.

> Leadership is not taught. It is modeled.
> —Dr. Emily Taylor,
> Former Dean of Women at the University of Kansas

How do you model leadership? Why should you model leadership? This book discusses the how; the why is because without leadership, employees will develop motivational dysfunction. What is motivational dysfunction? You see it everyday in retail operations and in businesses around the world. I have been asked, "If it is so common, why have I not heard of this before?" You have seen it, you have maybe even experienced it, and you have certainly seen the effects of this dysfunction in numerous operations.

Working with large and small companies, I have discovered that some companies' employees suffer from a previously undiagnosed dysfunction. Motivational dysfunction is that lack of motivation that employees seem to have due to a lack of motivation or excitement for their jobs. Reports put the loss to companies in the United States alone at over $3 billion per year from employees wasting time at work — gossiping, surfing the Internet, taking long breaks, and other such activities that do not add value to the company or the customer. Motivational dysfunction is a corporate and individual disease caused by a lack of leadership and direction in an organization. I have watched previously highly motivated employees succumb to motivational dysfunction with a change in leadership team of an organization. A classic example of this is the morale of the employees of The Home Depot when Bernie Marcus and Arthur Blanks retired and handed the reins to Bob Nardelli. As a soldier I saw this phenomenon happen to units when they had the ceremonial change of command. As a consultant I have seen this dysfunction appear in companies when a good leader is promoted or retired and a new "leader" takes over the reins. I have even seen motivational dysfunction appear in classrooms when students encounter a new professor/teacher who is not concerned with whether or not the students learn anything from the class.

Psychologists and psychiatrists have diagnosed many dysfunctions; and the Sprint Corporation, in a television advertisement debuting on Super Bowl Sunday, even discovered *connectile dysfunction*. I do not pretend to be a psychologist or psychiatrist. In fact, in addition to an introductory undergraduate course in psychology, the closest I came to being a psychologist was dating a psychology major for a couple of years. Motivational dysfunction does not require a trained medical specialist to cure it; what it does require is a world-class leader to provide the motivation and direction necessary to prevent or cure motivational dysfunction (MD). Here are a couple of good examples of personnel with MD:

1. Overheard in a large office building: "The Internet is down; what do I do now?"
2. Overheard in the halls of a FORTUNE 500 company: "She is having a terrible day with customers; I sure am glad that I am too busy to help her."

MD is a drain on public and private companies as well as governmental organizations. A report on Yahoo.com from Reuters in July 2007 based on a series of surveys showed that up to 34 percent of workers across companies admitted to wasting time surfing the Internet. Sixty percent of the respondents to the survey admitted to wasting almost two hours a day. In excess of 11 percent of the respondents listed the lack of challenging (motivating) work as the excuse for wasting time on the clock. This is a symptom of MD.

MD is not a medical condition that requires treatments or medicines. This dysfunction is a direct result of leadership and management within an organization. The cost to a company can be extremely high. The cure can be relatively inexpensive. The cure for MD is usually found in the foundations of leadership for the organization. That is where the root of the dysfunction is found. All too often, leadership and management are used interchangeably. The two concepts and terms are not the same. We look at this in greater detail while establishing the foundations of this study of leadership. Curing MD will set the conditions for success and excellence in your operations.

How do you set the conditions for success in your operations? As a soldier, I studied the history of my profession, the theorists and theories of warfare, and how to develop strategy and plans. I also studied the history of logistics, logistics theory and applications, and the importance of logistics in ensuring military operational success. These studies enabled me to be able to set the conditions for success in logistics and distribution operations from simple distribution to managing the distribution systems for Operation Iraqi Freedom.

As a leader and a coach, I studied the theories of management and the theories of leadership and how to effectively lead people from a wide variety of backgrounds in various operations throughout the world. The study of leadership principles and practices revealed how world-class and low-class leaders acted and reacted in certain situations. These studies did not teach me how to be a leader or coach but did show me how leaders and coaches respond to different situations. Observing leaders at all levels of organizations enabled me to develop the attributes of world-class leadership that form the "house of leadership." These studies and observations enabled me to set the conditions for success in operations.

Many of the FORTUNE© Best 100 Places to Work incorporate hours of training their personnel in the aspects of their professions. Why do they invest so much time and money in training? The investment of time and money in training enables the companies to set the conditions for success in their operations.

In warfare, it is important to set the conditions for success early in the planning and execution phases to ensure operational success. In any business, it is just as important to set the conditions for success. Regardless of the business that you think you are in, you are in the people business. Therefore, to set the conditions for success in any business endeavor, you have to set the foundation of leadership to ensure success.

In 500 BC, Sun Tzu wrote in *The Art of War* that "leaders are the safeguard of the Nation. When this support is in place, the nation will be strong. When this support is not in place, the nation will certainly not be strong." Substitute your company name for the nation and this translates to: "Leaders are the safeguard of the Company. When leadership support is in place, the company will be strong. When the leadership support is weak or not in place, the company will be weak."

In twenty-first century supply chains, if leadership is weak, the company will not be a leader and may find itself out of business. As more and more companies move toward using third-party logistics (3PL) providers[1] those companies are acknowledging that supply chain management is not one of their core competencies. It takes a strong leader to make the decision to contract out functions that have traditionally been in-house. Contracting out supply chain functions is not a new trend. The U.S. Army has used contractors to support the Army since the earliest days of the Army. General Washington used contractors to support his Continental Army and this trend has continued through present times.

Leadership skills are necessary when dealing with 3PL providers. And this is not a new problem. Alexander the Great covered the European, African, and Asian continents with purchased manpower as 3PL providers for his Macedonian Army. In the United States, centuries later, General Andrew Jackson learned the same lessons during the War of 1812. A look at his military achievements by the U.S. Army's Center for Military History reveals that:

> His force of will was decisive because the obstacles working against him were legion; he faced an elusive and adroit enemy; his soldiers were on the verge of mutiny; his supply system did not work...[2]
>
> From the outset, the supply contractors had experienced difficulties in meeting the terms of their contract. ... Jackson was forced to change contractors repeatedly. Each one, in turn, failed to get supplies to the army. ... General Jackson firmly believed that, unless an effective supply system could be established, "a pretext will be given for sedition, mutiny, and desertion, as has heretofore arisen, and which has destroyed the best of armies in the world...."[3]

Leadership by General Jackson was the key to solving these contractor problems and keeping his soldiers' morale high at the same time.

More than a few business leaders, starting with Peter Drucker in the early 1980s and more recently Thomas Friedman in his acclaimed book, *The World Is Flat,* have insinuated that the supply chain is where future competitions will be. Friedman states that "The 21st century will be defined by global competition and excess supply." Dr. Eli Goldratt proclaimed at an APICS International Symposium presentation a few years ago that "future competitions will be among supply chains." Competitions require leadership for companies to be successful.

Former Speaker of the House Tip O'Neill was famous for the statement that "all politics are local." All supply chains are local from the customer's perspective. As the world flattens and supply chains lengthen and become more complex, leadership of the supply chain personnel will continue to become more important to win the competition between supply chains. Leading the people in your supply chain will determine if your supply chain is one of global competition or one of excess supply. The choice really is yours.

With competition being between supply chains in the twenty-first century, we do not have the luxury of keeping a contractor that is not performing. Leaders need to understand what General Jackson understood almost two hundred years

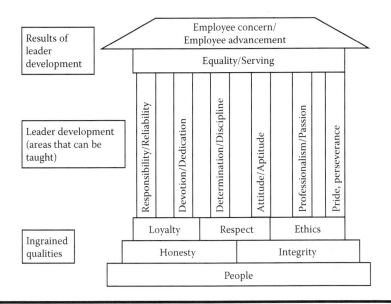

Figure F.1   The House of Leadership.

ago — without effective supply chain operations, companies will not be successful and the desertion that General Jackson was concerned about will be the desertion of customers to companies with effective supply chains.

This book looks at the attributes of effective leaders using the LEADERSHIP© acronym to establish the framework for developing leadership skills in yourself and your subordinates. We develop the framework using each letter of LEADERSHIP© as the title of the chapters. Each chapter looks at the critical attributes by defining the attributes and using examples of good and bad leadership from the military and commercial industry. The study of operations management uses the House of Quality to describe how quality is designed into and built into a product; this book looks at the attributes of leadership to form the House of Leadership. The House of Leadership using these attributes looks like that in Figure F.1.

This book provides a guide to building this house in yourself and your employees. The foundation of the house is people. Regardless of the business that you think you are in, you are in the people business — this is why people form the foundation of the House of Leadership and not because too many individuals want to walk over people to get to the top. The foundation of leadership is people; without people to motivate, inspire, and provide direction to, there is no need for leadership.

The critical attributes provide a framework to focus your leadership development programs and enable you to set the conditions for success in any operation. The attributes within the House of Leadership will also provide you with a framework for benchmarking your own professional leadership style and personal professional development program.

Recent articles in the *APICS Magazine* and in the *Council of Supply Chain Management Professionals* (*CSCMP*) *Journal* insinuate that there may be some real confusion as to what a supply chain leader really is. The *APICS* cover article[4] would have one believe that only senior executives are leaders, and the *CSCMP* article on its Web site would lead one to believe the same thing. It is important that we do not define leadership as only a senior executive trait or only in the boardroom realm. Other recent articles and presentations would have one believe that a supply chain leader is a company that is the leader in the industry. This may be true to some extent; however, to become a world-class company or leader in your industry, you must have world-class leaders and should be able to grow those leaders in your organization by modeling the correct attributes and actions for your future leaders to emulate.

Leadership is, or at least should be, at all levels in an organization. There is the defined formal leadership and the informal leadership in organizations that work together to create excellence in an organization. In an article entitled "Leaders Make Values Visible,"[5] Marshall Goldsmith states that:

> No business or strategy is good enough to succeed without strong leadership.... Companies that do the best job of living up to their values recognize that the real cause of success — or failure — is always the people...

Although this book is entitled *Modeling and Benchmarking Supply Chain Leadership,* the focus of the book is not about leading supply chains — it is about leading the people who make up the supply chain organizations. Do not make the classic mistake of thinking that you lead the supply chain organization — you **lead the people** who make up the organization and you **lead the people** who lead people in your organization.

What are you modeling for your subordinates? Are you modeling ego and self-promotion? A true leader does not seek the limelight for him/herself. He/she does seek to put his/her subordinates in the limelight. One major organization had a leader who became accustomed to being in the limelight as a result of a job as the chief spokesperson for the company. When he was promoted to a new position, he continued to seek the limelight rather than seeking to focus the spotlight on his subordinates. Leadership is not about your own personal ego. Seek to focus the spotlight on your subordinates and seek to promote your subordinates. In the long run, promoting your subordinates will prove to be the legacy that you leave behind on your company.

In describing Dwight D. Eisenhower in his book, *Eisenhower on Leadership,* Alan Axelrod states, "His task was not to lead men into battle but to lead those who led men into battle." This is equivalent to the premise that a leader does not lead an organization but instead leads the people who make up that organization. A supply chain leader does not lead the supply chain of a company; he or she leads the people who make up that supply chain. The people are the product, the movement of goods and services is a by-product of the leadership of the supply chain.

The cure for motivational dysfunction in your supply chain is world-class leadership. *Modeling Supply Chain Leadership* will provide the framework to lead the people in your supply chain to achieve new levels of excellence by helping you set the conditions for supply chain victory. Use these principles as a guide to help you "dare to inspire the people at all levels of your organization to reach new levels of personal and professional excellence every day!"

In twenty-first century supply chains, if the majority of the sessions at all the major supply chain professional organizations' conferences are an indicator, there is way too much emphasis on the technical skills, the automation of supply chain functions, and the information systems aspects of supply chain management. *Modeling Supply Chain Leadership* focuses on the people aspects of our profession. Without the right, trained, and competent people, our supply chains will fail. And when people are involved, the critical aspect is leadership. Your supply chain employees and supply chain partners are looking to you for an example and leadership. Give it to them!

Notes

1. In his book *Supply Chain Best Practices*, Dave Blanchard suggests that more than 63 percent of companies are outsourcing some aspect or all of their supply chain operations.
2. Medley, James E., LTC. 1998. Studies in Battle Command, Andrew Jackson's Iron Will in the Creek War, 1813–1814, Combat Studies Institute, Fort Leavenworth, KS, p. 21.
3. Medley, James E., LTC. 1998. Studies in Battle Command, Andrew Jackson's Iron Will in the Creek War, 1813–1814, Combat Studies Institute, Fort Leavenworth, KS, p. 22.
4. *APICS – The Performance Objective Magazine*, December 2007.
5. Goldsmith, Marshall, Leaders Make Values Visible, www.MarshallGoldsmithLibrary.com. Accessed January 8, 2008.

Preface

Setting the conditions for success — what does that mean? In military operations it is important to lay the groundwork, develop the plans, develop the operational requirements, and prepare for any shifts in operational plans. Failure to do so will result in failed operations, loss of confidence in leaders, and the potential loss of lives. In leader development, it is just as critical to set the conditions for success in an operation. Sun Tzu wrote that leaders are the stewards of the country. He went on to say that when the leader is strong, the country is strong; and when the leader is weak, the country is weak. Substituting company for country, the new translation of Sun Tzu's *The Art of War* would state that leaders are the stewards of the company; and when the leader is strong, the company is strong. To develop strong leaders, the conditions must be set for success by having a strong leader development program.

The House of Leadership and the critical success indicators of world-class leadership set the conditions for success for your operations. It is vital to remember that leaders do not lead organizations, companies, or even departments; leaders lead the people who make up these organizations, companies, and departments.

A recent leading operations management and supply chain magazine had a lead article that promised to discuss supply chain leadership. The article had some good points but left the reader with the impression that only senior-level personnel are supply chain leaders. This article also insinuated that supply chain leadership is a career path toward advancement in an organization. Leadership is about motivating people, giving the employees a sense of purpose, and successfully meeting the objectives and missions of the company. It is not about career advancement. Progression and advancement are collateral benefits but not the purpose of leadership.

In a presentation in early 2002 for the *World Supply Chain Summit*, the Supply Chain Leadership Institute proposed that supply chain leadership was the next step in the evolution of supply chains as we continue to evolve from pure logistics functions to supply chain management, to supply chain synthesis, and now to supply chain leadership. Today there are graduate- and undergraduate-level programs that focus on supply chain leadership. This book assists individuals and companies in developing supply chain leaders and helps those companies model and benchmark leadership throughout their organizations. And by developing supply chain leaders who effectively lead the employees who make up the supply chain, these supply chain leaders will become enablers in creating companies that are leaders in their supply chains.

About the Author

Joseph L. Walden (Colonel, U.S. Army, Retired) has more than thirty years of leadership and supply chain experience as a practitioner (the *Supply and Demand Chain Executive Magazine*'s 2004 Supply Chain Practitioner of the Year; 2003 Top 20 Logistics Executives in America), as a consultant, and as an educator at the undergraduate and graduate levels. His experience includes designing and operating the multi-million square foot distribution center in support of Operation Iraqi Freedom and consulting for FORTUNE 500 companies as well as the Department of Defense.

As a coach, he coached the 1989 Women's National Powerlifting Team and served as coach of the Armed Forces and All-Army Powerlifting Teams from 1984 to 1990. The 1984 and 1990 teams also won the national championships. As a competitor, he set more than 70 national and state records in the 181- and 198-pound classes in powerlifting.

Walden has written numerous articles for national and international lifting magazines and for national and international supply chain magazines.

Walden retired from the U.S. Army as a Colonel and now serves as the Executive Director for the Supply Chain Leadership Institute and the Midwest Leadership and Coaching Center. He is the Director of Education for the Warehousing Education and Research Council and an APICS Certified Fellow in Production and Inventory Management. He serves as an assistant professor of operations management and logistics for Webster University and a supply chain lecturer for the University of Kansas.

SETTING THE FOUNDATION

I

The goal of this section is to establish a clear definition of *management* and a clear definition of *leadership* — what they are and what they are not. By establishing these definitions, we can then make a comparison of the similarities and differences between management and leadership and why both are necessary while emphasizing the need for leadership to ensure a world-class organization made up of world-class people.

The *foundation* of any organization, regardless of size, is the people who make up the organization. In building a house, the footings and foundation must be set and cured before the house can be built. The definition of *leadership* is the footings, and the people are the foundation. Just as the footings and the foundation are the first steps in building a house, *Modeling Leadership* will start here also.

After establishing the foundations of leadership, Chapter 2 looks at the teachings of Sun Tzu from his seminal work *The Art of War* as it relates to leadership. Sun Tzu will help in establishing a foundation in leadership before addressing the attributes of world-class leaders in Section 2.

Chapter 3 takes the concept of Six Sigma and applies it to leadership. Six Sigma Leadership is not about leading Six Sigma projects. That is best left for the Black Belts and Master Black Belts. This application of the Define-Measure-Analyze-Improve-Control concept looks at the development of leaders in any organization using the principles of Six Sigma. It is not about zero defects leadership — that will never exist because there are no perfect leaders. Like the Sun Tzu applications, the application of Six Sigma to leadership will form the foundation for looking at the attributes and enablers that world-class leaders model for their supply chain employees.

Chapter 1

What Is Leadership, and How Does It Differ from Management?

> Team members at all levels of an organization need to understand what leadership is and what leaders do.

All too often, the terms "leadership" and "management" are used synonymously and interchangeably. Are they really interchangeable terms, or are they separate disciplines? Many companies in their annual reports address the senior management team — this usually includes the president and vice presidents of the firm. Are these managers or leaders? Mary Kay Cosmetics is one of the pioneers in calling its senior leaders the Leadership Team. Can you be both a leader and a manager? This chapter establishes the differences and similarities between leadership and management, and also establishes workable definitions of both to assist you in breaking out the differences in your company.

As of the date this was written, there were 482,589 books on management and 207,481 books on leadership. In the supply chain field there were more than 12,900 books on supply chain management and only 21 books that addressed supply chain leadership — a concept first introduced by the Supply Chain Leadership Institute in 2001. MBA programs look at the functions of managers but rarely address leadership in the education of future "leaders of business."

Every organization, regardless of size, has a requirement for both leaders and managers, and in small companies there may be one person doing both of them. Every leader has some managerial functions but not all managers have leadership functions

3

in their job descriptions. Not everyone can be a leader, and not everyone wants to be a leader. There are people who are content to be followers only. Even leaders must sometimes be followers to be effective. Everyone has someone to whom they report.

Ask a hundred people to define leadership and you will get a hundred different answers. It would lead you to believe that perhaps leadership is a lot like love — everybody knows what love is when they experience it but they cannot necessarily define it. Everybody has experienced good and bad leadership and can recognize it when they see it or experience it but they cannot necessarily define it. One of America's greatest World War II leaders, General George S. Patton, Jr., could not define it but did know "that it is the thing that wins wars." General Dwight D. Eisenhower defined leadership on the day before the Normandy Invasion as "the art of getting someone else to do something you want done because he wants to do it."

We are all in wars every day that require leadership. Some wars are military and political in nature and the need for leadership is evident and the lack of leadership can result in the loss of lives and, in all too many cases in history, the loss of a country. In business we face war-time situations every day. Your competition is working hard to take your business away, and the loss of business can and often does lead to the loss of the company. We are also in a war against technology — if we lose the war of technology, our competition jumps ahead of us in the marketplace and the potential of losing the company is indeed possible.

Webster's Dictionary provides the following definitions:

- Lead: To guide or direct in a course.
- Leader: One who leads or guides; one who is in command of others.
- Leadership: capacity or ability to lead.

Unfortunately, none of these definitions provides a good foundation for the discussion of leadership. Perhaps the best definition of leadership is found in the United States Army Field Manual, *Leadership*.[1] This Field Manual defines leadership as "The act of influencing people by providing purpose, direction, and motivation; while operating to accomplish the mission and improve the organization."[2] This will serve as our definition of leadership throughout this book. To further develop

Definitions
• Lead: To guide or direct in a course
• Leader: One that leads or guides; one who is in command of others
• Leadership: Capacity or ability to lead

Figure 1.1 Leadership definitions to help set the foundation.

the foundation in order to start setting the conditions for success in any operation, let's break down this definition into workable phrases.

The U.S. Air Force doctrinal manual on leadership defines leadership as "the art and science of influencing and directing people to accomplish the assigned mission. This highlights two fundamental elements of leadership: (1) the mission, objective, or task to be accomplished, and (2) the people who accomplish it. ... The leader's primary responsibility is to motivate and direct *people* to carry out the unit's mission successfully. A leader must never forget the importance of the *personnel* themselves to that mission."[3]

The U.S. Air Force definition is not that much different from the U.S. Army's definition but the second part of the introduction to the chapter on leadership identifies the critical aspect that must never be overlooked, regardless of what your mission or objective is, and that is the people who accomplish the goals, missions, or objectives of your organization. The goal of leadership is therefore to transform the potential of the people in the organization into effective performance. A secondary goal of leadership is to develop leaders for the future of the company.

> The inspiration of a noble cause involving human interests wide and far, enables men to do things they did not dream themselves capable of before, and which they were not capable of alone.[4]

The Act of Influencing People

Exactly what is the act of influencing people? Influencing people is what General Eisenhower was referring to when he spoke of getting someone else to do something because *they want to do it.* Influencing is getting employees to do what is necessary. Influencing employees goes past just giving out assignments. Setting the example is discussed in Section II. This is an important method of influencing others. How many bosses have you had who influenced your actions — good or bad? Jackie Robinson may have hit this square on the head when he said, "A life is not important except in the impact it has on others' lives."

Do your actions have an impact or influence on the lives of others? The answer to this question is yes unless you live alone, work alone, and never venture into the real world. And even then, your actions have impacts on others, just not necessarily in a positive way.

How do you influence people? Remember that whatever business you are in, you need people to accomplish your objectives. Jackie Robinson said that the value of a person's life is measured by the impact he has on other people. As a leader, you have an impact on a large number of people through your actions. We will look at the impact of your example (good or bad) when we discuss the attributes of world-class leadership. One of my sister units when I was a lieutenant had a leader who had so much influence through his actions that his officers started talking just like him.

As a leader, your actions, your mannerisms, your life style, and your words have a big influence on those around you. As a leader, you have to establish a clear vision for your organization and for the individuals who make up that organization. Your vision for your organization should include the goal of the organization. What is the goal of your organization? Can you explain it in a few sentences? When I was with the U.S. Army Velocity Management Program,[5] we had what we called our "elevator card." This card had the mission and vision of Velocity Management on a three-by-five-inch index card and provided a few key statements of the goals, vision, and mission. Your vision should be able to be reduced to an elevator card. If you cannot describe what you are doing or where you are going in a few sentences, then your vision is too complicated. Dr. W. Edwards Deming said it a little differently when he said, "If you cannot describe what you are doing as a system, you do not know what you are doing." As a leader setting the vision for your organization, you have to know what you are doing, where you want the company to go, and where it is actually going.

This vision must be tied to the organizational goals and strategy and may in fact influence the goals and strategy of the organization. Your vision cannot run counter to the organization. If you are a retail organization, your vision must incorporate customer care, customer service, and customer satisfaction to ensure continued profitability. The vision must be clearly stated, clearly articulated, and clearly understood by the employees who have to implement the actions to make the vision successful. Stephen Covey called this "seek first to understand and then be understood."[6]

Making the vision clearly understood is necessary, and it must be understood from the perspective of the people who have to implement it. Each business has its own language. In the Army we had a language that seemed foreign to people outside the military; in fact, we had such a foreign language that it was sometimes not even understood inside the Army — this is not a good thing when the same words or acronyms mean different things to different people in the same organization.[7] In the supply chain industries there is a language that is foreign to people outside the supply chain. In manufacturing there is "foreign" language — even the food service industry has a language all its own. The key is to ensure that the vision is clearly understood by everyone in the organization; therefore, it must be in a clearly understood language and stated in such a way that the people who have to implement it not only understand the vision, but also can translate the vision into the language of their favorite radio station — WIIFM - "What's In It For Me?". This is the first step in influencing people.

I have discovered a new phenomenon in the corporate world that did not exist in the military. It is relatively easy to motivate soldiers who are getting shot at or may face the opportunity to get shot at in the future. Protecting yourself and those around you is a great motivator. But in corporate America I have discovered what I call *motivational dysfunction* (MD). The cure for MD is a clearly stated vision and leadership that provides purpose, direction, and motivation to the entire workforce.

Purpose, Direction, Motivation

What is your purpose? How do you convey it to your co-workers and subordinates? What is the purpose — the driving goal or core competency — of your organization? Is your purpose in line with the purpose of the organization? The vision sets the foundation for the organization. However, the real core competency or prime directive of your employees is what motivates them. As the leader, it is your responsibility to convey your vision and the vision of the organization in such a way that it becomes the motivator or vision of the employees. Again this ties back to WIIFM. Each individual employee is motivated differently. This complicates the leadership equation because this means that there is no one-size-fits-all style of leadership. You must be flexible as a leader to adapt your style to the needs of the employee. One way of doing this is to ensure that every employee knows what is needed, when it is needed, and why it is needed, as well as how he or she fits into the overall success of the firm. After the completion of the American Civil War, one of General Grant's aides stated that what made Grant so successful was that "he made sure that all of his subordinates knew exactly what he wanted, when he wanted it and why he wanted it."[8]

While establishing the Theater Distribution Center for Operation Iraqi Freedom in Kuwait, I thought I had made it very clear what our purpose was. To my surprise when I asked the soldiers assigned to the Center their purpose, I was told "to move stuff from one truck to the customer lanes to the outbound trucks; and to drive those forklifts." Ask your forklift drivers the same question and you will probably get a very similar answer.

I gathered all the soldiers together and explained to them the importance of the Theater Distribution Center in the overall success of Operation Iraqi Freedom. I explained to each of them that they were the last link between the wholesale supply systems in the United States and Germany, and the soldiers going into combat. I tried to make it very clear that unless we did our jobs correctly with accuracy and precision, there would be soldiers who would be without food, supplies, and ammunition. I told the soldiers that they were the most important link in the supply chain as the interface between the supply system and the customer. After that little discussion, their chests puffed out and they went about their work with true purpose and motivation.

One of my favorite questions for distribution center personnel is to ask them what their job is. The purpose of the question has several functions. The first is to find out exactly who is doing what in a distribution center. The second function of this question is to find out how motivated the employee really is. Are they there for a paycheck, or are they there because they enjoy their jobs and feel like they are contributing? If they answer like my forklift drivers did — and most do — let them know that they are an important part of the supply chain and may very well be the last person in the company to touch a product before it goes to the customer. See if that does not change their outlook and disposition.

Every worker in any environment needs to have a purpose that they understand in clear, concise language; they need a direction to point toward and a reason for working. Have you ever seen a compass that has been demagnetized? The arrow of the compass spins and points in every direction except North. Workers are the same way; they need leadership to provide them with direction and keep them from becoming demagnetized or suffering from motivational dysfunction.

Unfortunately, every worker is not motivated in the same way. What motivates workers? What motivates workers is what motivates each individual worker. There is not a one-size-fits-all motivation — that is what makes motivational dysfunction an interesting malady. The House of Leadership sets the foundation and structure for developing leaders who understand employees, who understand employee needs, and who can provide the purpose, direction, and motivation to prevent the onset of motivational dysfunction and help companies and employees cure this growing phenomenon in the corporate world.

Accomplishing the Mission

All too often, managers and leaders become so consumed in accomplishing the mission that nothing else matters. Unfortunately, this drive to accomplish the mission is because of the drive to climb the ladder of success. "If I can get this mission accomplished, I will be promoted." This drive leads to motivational dysfunction for workers because they do not share the same goal to climb the ladder or because they do not see the same benefits as the driven manager or leader.

There is another drawback to the drive to accomplish the mission — what is the next challenge, what are you sacrificing to get the mission accomplished, and is the satisfaction of accomplishing the mission going to keep you motivated for the next mission? Leaders at all levels, throughout history, have sacrificed their personal lives to climb the ladder. There was a song a few years ago that asked the question, "Will there be a gravestone with the words — 'If I could have only spent a few more hours in the office'?" Corporate leaders and military leaders have all too often sacrificed their families for career success. My experience tells me that time away from the family is necessary but there is a balance that is addressed by the House of Leadership — because time lost with your family can never be replaced.

Accomplishing the mission is important for the success of the organization. The key is to ensure that the missions that you are working on are critical to the success of the organization and not just your pet project.

And Improving the Organization

You have to ask yourself if the goals you are setting and the missions you are trying to accomplish are good for the organization, or if they are just good for your career. This is the key to establishing quality leadership. Moving up the corporate ladder is a collateral benefit of accomplishing the mission and improving the organization. Corporate America is full of "leaders" who have climbed the ladder on pet projects that have improved their careers but were suboptimal to the overall success of the company. Today's focus on the bottom line as the only measure of success leads to short-term thinking versus long-term goals to improve the organization. "If I can make things better during my tenure and then move on, I really do not care what happens after my promotion." This is an attitude that, although not always specifically stated, has been implied by way too many corporate "leaders."

One key method of improving the organization is through the use of developmental counseling and establishing training programs for employees. Developmental counseling is a critical program that enables leaders to help their employees improve their performance by explaining what the employee is doing wrong and how to improve in those areas; this may very well include performance improvement training or retraining. Developmental counseling also includes what the employee is doing right. All too often, counseling programs focus on the bad and neglect what the employee is doing right, and therefore has a tendency to demotivate the employee and is seen as a negative program.

The definition of leadership is so comprehensive that it must be stated one more time as the foundation for the discussion of modeling supply chain leadership:

> Leadership is the act of influencing people by providing purpose, direction, and motivation; while operating to accomplish the mission and improve the organization.

This form of leadership will set the conditions for success in any operation! The next chapter looks to Sun Tzu and his book *The Art of War*. Sun Tzu provides the historical foundation of leadership prior to a detailed look at modeling supply chain leadership.

Notes

1. *Field Manual 5-0, Leadership,* U.S. Army. The United States Army uses Field Manuals as doctrinal publications. This particular Field Manual is devoted to leadership development and leadership skills needed for any U.S. Army operation.
2. *Field Manual 5-0*, Leadership, U.S. Army, 2006.
3. *U.S. Air Force Doctrinal Manual,* p. 1-1.
4. Joshua Chamberlain, speaking in 1889 at the dedication of a monument to honor/memorialize the 20th Maine Regiment's actions at Gettysburg on July 2, 1863. General Chamberlain went on to receive the Congressional Medal of Honor for his actions on the battlefield that day. General Chamberlain was later present at the surrender of General Robert E. Lee at Appomattox Court House in 1865. As General Lee passed, General Chamberlain showed his adversary respect by bringing his soldiers to attention and saluting the Confederate general. This action received ridicule from other officers but was a demonstration of General Chamberlain's character and his respect.

5. The Velocity Management Program served as the U.S. Army's Supply Chain Process Improvement Program. The program started by looking at customer wait times and order cycle times and expanded to include financial impacts and constraints, a close look at what was stocked where and in what quantities and the impacts of maintenance operations on the supply chain. This program later looked at the reverse logistics processes and drivers long before reverse logistics became a hot topic in supply chain discussions. The Velocity Management Program was loosely based on Six Sigma methodologies but in lieu of the traditional Define, Measure, Analyze, Improve, and Control (DMAIC) framework, Velocity Management used a RAND Arroyo Center designed methodology — Define, Measure, and Improve (DMI). All the traditional DMAIC tools and concepts were encompassed in DMI.
6. See Covey, Stephen R. 1989. *The 7 Habits of Highly Effective People,* FranklinCovey, Salt Lake City, UT.
7. I have since learned that many corporations have this same problem. Acronyms and definitions for words mean different things to different departments. This causes great confusion and sometimes results in demotivating employees or missed shipments — neither of which is good for any company.
8. General U.S. Grant was separated from the U.S. Army prior to the start of the American Civil War because of his drinking problems while stationed in California, then failed in business in Illinois before being called back to duty in the militia, and then went on to become the most successful general for the U.S. Army, bringing about the fall of Vicksburg, the fall of Richmond and Petersburg, and the defeat of the Army of the Confederate States of America when he accepted the surrender of General Robert E. Lee.

Chapter 1 Questions

1. Is leadership different from management even though they are frequently used interchangeably?
2. Am I applying the definition of leadership to my organization?
3. Am I leading my people, or am I trying to manage them as assets?

Chapter 2

Sun Tzu on Leadership

In this work you will learn how people are to be treated and dealt with.
—Stephen Kaufman[1] describing The Art of War in his translation of
Sun Tzu's seminal treatise on strategy

What can we learn from a military theorist from 500 BC about leadership in the
twenty-first century? And just who is Sun Tzu? And just what does some dude from
500 BC China know about supply chain leadership? *The Art of War* continues to be
required reading for a large number of our trading partners and supply chain partners in the Asia-Pacific Rim. This chapter takes a quick look at Sun Tzu, his writing,
and the applications of this ancient book on strategy to leadership and supply chains
to form the foundation for our discussions on modeling supply chain leadership.

There are more than twenty quality translations of *The Art of War*. Some of
these translations ponder the idea of Sun Tzu's existence at all. Some postulate
that *The Art of War* is a compilation of several writers. One translator even went
as far as suggesting that Sun Tzu was really Lao Tzu, the author of the *Tao Te
Ching;* another translator suggested that Sun Tzu was a younger contemporary of
Confucius. Archeological finds in the past twenty years point to the fact that Sun
Tzu did indeed live in the Wu Province of China around 500 BC.

Sun Tzu wrote *The Art of War* as a living document that detailed his thoughts
on military operations. Because of the way he wrote the short treatise, it can be
adapted to more than just military operations. The first translation of *The Art of
War* in the Western World was not until around 1796. This translation was in
French. It is no small coincidence that the first translation was in France and that
the tactics of Napoleon reflected the theories of Sun Tzu.

Reportedly, the study of Sun Tzu was part of the education process for the officers of the Third Reich and some of their operations, to include the attack that became known as the Battle of the Bulge, demonstrate Sun Tzu's ideologies.

Mao Tse-tung based the works of his *Little Red Book* on the writings of Sun Tzu. Ho Chi Mihn and the Viet Cong used principles and tactics reminiscent of Sun Tzu's writings in the war between North and South Viet Nam. A good example of this was the Tet Offensive — striking when unexpected at a perceived weak point.

The first translation into English did not occur until Lionel Giles translated *The Art of War* in 1910. This translation serves as the basis for most available translations today. The U.S. military studies Sun Tzu as part of the Command and Staff Colleges' curricula and at the War College[2] level. Interestingly, the Air Force Command and Staff College and the Air Force War College place more emphasis on Sun Tzu, and the U.S. Army places emphasis on the study of Karl von Clausewitz. Napoleon based his theories of warfare and leadership on his study of Sun Tzu, and Clausewitz developed his theories of warfare and leadership based on his study of Napoleon. The writings of Clausewitz, based on his observations of Napoleon, influenced U.S. Civil War leaders such as William T. Sherman and continue to influence U.S. military theories today.

For this look at Sun Tzu and leadership, we use direct quotes from passages as well as take some of his passages and substitute leadership for warfare and for generals. Using this substitution, the first passage of *The Art of War* states, "1. Sun Tzu said: The art of war is of vital importance to the State. 2. It is a matter of life and death, a road either to safety or to ruin. Hence it is a subject of inquiry which can on no account be neglected."[3] This opening line becomes "Leadership is of vital importance to the Company. Leadership is a matter of life and death for the company, the road to either success or ruin. Therefore leadership is a subject of inquiry which can on no account be neglected." For this reason we look at some of the other passages from *The Art of War* for establishing a historical foundation for leadership as a cure for motivational dysfunction (MD).

Who was Sun Tzu, and what is his relevance to a discussion of leadership and curing MD? Let's start with who he was and then look at the relevance of his writings to a detailed look at leadership as the cure for MD.

According to the Web manager for the Web site http://www.sonshi.com, Sun Tzu is literally translated to mean Master Sun.[4] Most historians agree that Sun Tzu lived in the Wu Province of China in the late 500 BC period. Sun Tzu came to prominence when one of his acquaintances introduced him to the King of the Wu Province as a military strategist who lived as somewhat of a recluse. Sun Tzu wrote his short treatise *The Art of War* as a guide to military strategy.

Sun Tzu told the king that he could demonstrate the effectiveness of his theories and tactics. The king assembled a room full of ladies from the kingdom. Sun Tzu instructed the ladies in the art of drill and ceremony. When he was ready to show the king that anyone could learn his techniques, Sun Tzu put two of the king's favorite ladies in charge. He asked them if they understood the directions; they said yes. But when he gave the orders, the ladies laughed. He asked them again if they understood the training and the commands; again they said yes. Sun Tzu was demonstrating what is today better known as one of Stephen Covey's "7 Habits" — the habit of seeking to understand before being understood. Sun Tzu told the king that if the instructions and commands were not clear, then it was the fault of the leader; if they were understood and not followed, then it was the fault of the followers.

His next action is not recommended when trying to gain favor with the boss. Sun Tzu had the king's two favorite ladies beheaded.[5] Sun Tzu then put two more ladies in charge, asked again if they understood the commands, and then had the room full of ladies demonstrate the drill and ceremony movements with military precision.

Leadership and Success

Sun Tzu stated in Chapter 1 of *The Art of War* that there are five factors that must be mastered to be successful in operations. Leadership was one of those factors. In Chapter 2 he states, "Therefore, a general who understands warfare is the guardian of people's lives, and the ruler of the nation's security."[6] Paraphrasing this to today's business environment would have Sun Tzu saying that a leader who understands his/her business is the guardian of the life of the company and the people employed by the company.

Setting the Example

Sun Tzu also wrote of the importance of a leader not losing his temper and setting the example for his followers. Some 2500 years later this is still important. Leaders must set the example for their followers to emulate. The actions of leaders at all levels are on display to their employees and co-workers.

We will look more at the concept of setting the example in the discussions on the House of Leadership. Always remember that as a leader, as an adult, even as a parent, you are being watched by someone who is looking at you to see how to act, talk, or respond to situations. Every person who I have ever worked for, and in some cases worked with, has contributed to my theories on leadership. Some served as examples of how to act, speak, or respond; others have served as examples of how *not* to behave, respond, speak, or respond. After working for a couple of bosses, I have stated clearly that "I do not want to be like that if I get in a position of equal leadership." Fortunately, I have had more bosses who have positively impacted my outlook on leadership and have left me with a feeling of "I hope I can be like that when I get in their positions."

Know Yourself

In Chapter 3, perhaps the most famous of Sun Tzu's quotes is found: "One who knows the enemy and knows himself will not be in danger in a hundred battles."

In today's business this is just as important. A leader must know his/her capabilities, shortcomings, and opportunities to be successful. This is discussed in Section II when we look at modeling and benchmarking your leadership skills against the skills and attributes of world-class leaders.

How do you know what your capabilities are? What about your shortcomings? The components of the House of Leadership will provide you with a benchmark for your leadership and the leaders who work for you.

Another method of learning your capabilities and shortcomings is to perform a Strengths, Weaknesses, Opportunities, and Threats (SWOT) Analysis on yourself. Just as you would not venture into a new operation or product development without performing a SWOT[7] Analysis, you should do the same for yourself and your subordinates. In his interpretation of *The Art of War*, Stephen Kaufman[1] asks the question, "Do you believe in yourself as a leader?" Your personal SWOT Analysis will reveal the answer to this question. If you do not believe in yourself as a leader, you will never get your subordinates to believe in you as a leader.

Your belief in yourself as a leader is critical to your success as a leader. Your belief in yourself as a leader will lead to faith in your abilities not only by you, but also faith in your abilities from the people who you lead. Faith and belief in yourself are necessary if you are to be a successful leader. A classic example of faith and belief can be found in the Gospel of Mark. A father asked Jesus, "If you can, please heal my daughter. Jesus said to him, 'If you can believe all things are possible to him that believes.'"[8] This is true of your goals and visions but just as important in your belief in yourself as a leader. If you do not believe in yourself as a leader, how can you expect your employees or your employer to have faith in you as a leader?

Dealing with Employees

Sun Tzu wrote about how to deal with people — his dealing with the king's two favorite ladies is not a good example of how to deal with people. But Sun Tzu also wrote of the benevolence of the leader and the willingness of the people to follow a leader who is disciplined but fair. In Kaufman's interpretation from the martial arts perspective, he states that "The more humiliation you place on the enemy, the more vengeance he will crave...."[9] Although some managers want to be leaders and treat their employees as enemies (especially in union negotiations), never mistake your employees for enemies. However, the more humiliation you place on your employees, the more likely that your employees may indeed become your enemies. In today's society, technology enables disgruntled employees to steal confidential or proprietary data on a single eight-gigabyte memory stick or infect a large number of computers with an infected file.

About 500 years after Sun Tzu wrote about the need for leaders to be benevolent and fair while setting the example, Jesus gave his famous Sermon on the Mount. During this sermon, Jesus told the crown, "Treat others as you would like to be treated."[10] These words are known as the "Golden Rule." The Golden Rule is as true to leadership in the twenty-first century as it was when first said more than 2000 years ago. All too often, leaders believe that they can mistreat employees to raise production and get the leader promoted. And unfortunately, in a large number of cases, the leaders who follow the Golden Rule as a guide to leading people find themselves passed up for a promotion in favor of the person who leads through intimidation. However, almost every leader who I have encountered who used the Golden Rule as the guide to leading people would not change his/her style just to get promoted or advance in the company.

Metrics and Measures of Effectiveness

There are some consulting companies that will make you believe that they came up with the idea of developing benchmarks, metrics, and measures of effectiveness. Sun Tzu spoke of the need to develop measures of effectiveness for the organization and the need to have metrics and quality forecasts in order to have victory. Some 2500 years later, it is still important for a leader to be involved in the process of developing metrics to measure the quality of the company's output — regardless of the output of the company. The output could be retail service, products produced on a production line, or a third-party logistics support company. Regardless of the process and the products, the leader must be involved in the development of the measures of effectiveness and the quality metrics for the company.

Leading Small versus Large Organizations

In Chapter 5 of *The Art of War*, Sun Tzu makes it very clear that the principles of leadership are the same for a "small army" as they are for a "large army." This is true today. The size of the organization does not affect the principles of leadership that form the House of Leadership. Some organizations have the belief that a different leadership style is necessary at different levels of the organization. This may or may not be true. Sun Tzu tells us that the principles are the same. How a leader deploys the principles may differ at different levels of an organization because he/she may need different strengths in different leadership skills or may have to use some of the skills more often than other skills but I am sure you will discover that Sun Tzu is accurate in stating that the skills and principles are the same at any level of an organization.

Mergers and Acquisitions

How does your company deal with mergers and acquisitions? One Human Resources executive told me that there are no more mergers, only acquisitions. This may or may not be accurate, depending on your industry. However, the question remains the same: even if your company acquires another company, how do you handle the new employees? The Romans seemed to have the right idea. Their policy was to conquer a city-state/country and then offer Roman citizenship to the newly acquired citizens. Sun Tzu wrote of similar techniques from the other side of the world. He stated that the newly conquered soldiers should be assimilated into the formations and given the "colors" of the army.[11] Some would argue that there is a difference between the Romans conquering a city-state and a company buying another company. Observations of merged companies would indicate that we should listen to the Romans or Sun Tzu in today's businesses.

Here are a couple of examples. One FORTUNE 500© merged with another company of similar background and within the same industry. The larger of the two companies became the parent company. Almost two years after the "merger" into one company, there were still two corporate headquarters campuses. The corporate intranet contained a page for "Legacy X" employee travel and "Legacy Y" employee travel. There was a "Legacy X" finance link and a "Legacy Y" finance link. There were even "Legacy X" and "Legacy Y" benefits links on the intranet home page. At meetings, the employees spoke of "Legacy X" and "Legacy Y" technologies and referred to themselves as "Legacy X" or "Legacy Y" employees. The company continues to struggle with its image, its profitability, and its employees.

Did this company follow the guidelines of the Romans or Sun Tzu? It is pretty obvious that they did not. Would they have been more successful if they had followed the guidance of Sun Tzu to assimilate the employees into one company? Becoming one company is a challenge with which every executive of every company in a merger must deal. It is also a challenge for the employees of these companies. Leaders must make the employees of the new company feel part of the company quickly. Failure to assimilate the employees into one company will produce the same results every time, as was evident with "Legacy X" and "Legacy Y."

In a large government contracting company, the same results can be seen. Contrary to the teachings of Sun Tzu, this company merged two divisions into one larger, "more efficient" division that would be able to go after and get more business. The merger of the two divisions was done with great fanfare and promises of "one team." Some eighteen months after the fanfare and the announcement, there were still two sets of travel programs, two sets of retirement programs, and two sets of pay systems. The amazing thing here is that the old travel system of one division that was supposed to be rolled into the new division system is more expensive than the new corporate travel system but the employee travel vouchers for the old system employees are closed out in half the time compared to that of the "new division."

It is imperative to assimilate new employees into an organization as quickly as possible. Just as you would assimilate a new employee into an organization, you must also assimilate the employees of an acquired division or company into your organization and make them feel part of the corporate family with the same systems as everyone else as quickly as possible before their discontent is passed on to other employees.

Lead from the Front

During the American Civil War, it was common for the generals to be close to the front of the formations. After all, in those days, communications were a bit more complicated and complex than today. General Robert E. Lee set the example for his generals by leading from the front. In fact, his soldiers and subordinate leaders were sometimes concerned that he was too far forward and would be wounded or killed by being at the front. Many were heard to exclaim, "General Lee to the rear!" — not out of disrespect, but because General Lee was revered by his soldiers and they did not want to lose him.

During the epic battle at Gettysburg, Pennsylvania, being on the ground and leading from the front proved critical for both sides. The Confederate forces "led" by General Longstreet did not arrive at the proper place and time because General Longstreet was not convinced that the plan was a good one. This failure may be what enabled the U.S. Army to win the battle and preserve the country.

A relatively inexperienced schoolteacher and administrator on the other side of the battle modeled leadership for his soldiers. Colonel Joshua Chamberlain is credited with saving the day and the battle at Gettysburg because of his actions at Little Round Top. Because he was on the ground, Colonel Chamberlain could see what was happening on the battlefield and was able to assemble his soldiers at the end of the Union line and prevent a flanking action by the Confederate States Army. It is sometimes easier to motivate soldiers when they are being fired at than it is to motivate supply chain employees — but the principles are the same. Leading from the front is important, and you must be out of the office to know what is happening.

Sun Tzu stated that leaders must lead from the front and must get on the ground to see what is going on. General George S. Patton, Jr., said that "no effective decision was ever made from the seat of a swivel chair." Both Sun Tzu and Patton are saying the same thing: you have to see the operation for yourself to know what is really happening — no filters, no interpretations by subordinates.

In business, the common misconception is that a leader can lead from behind the computer screen. "He does not need people skills; he can see what is going on from the reports." How many times have you heard that statement? You can be a good manager from behind the computer screen because you can see the figures and make decisions based on the information in the computer or within the spreadsheets. However, because leadership is the act of influencing people, you must be with the people in order to influence them.

Here is an example of not getting out of the office. In one assignment, I was returning to a location that I had been assigned to just a few years earlier. My predecessor was adamant about showing me around the operations, even though I was familiar with them. In every military organization, there is a "Chain of Command" Board that has pictures of the Commander in Chief (the President of the United States) and all of the commanders and command sergeants major in the chain of command down to the company commander. In this particular case, my predecessor's picture was therefore in every company in the organization.

As we toured the facilities we came across a couple of soldiers who were doing their job in an unsafe manner. My predecessor asked them, "Do you know who I am?" I guess this was supposed to make a difference in whether or not what they were doing was safe. I was thinking that it did not matter who he was, but rather that they should follow safety practices. One of the soldiers scratched his chin and said, "I've seen your picture somewhere." It was all I could do to keep from laughing. The sad thing is that my predecessor had been there for two years and he did not know his employees, and his employees did not know him. So, if for no other reason, get out of the office so the employees know who you are.

Evaluate Your Performance

Sun Tzu tells us to evaluate our performance to improve our operations. He was concerned about learning from the operations so that the enemy did not assume his armies would do the same things again. In today's business, it is still true to evaluate our operations to ensure that we learn from our successes and our shortcomings. In the U.S. Army, after every training event, mission, or operation, the unit conducts an After Action Review (AAR).

The AAR is a great tool for evaluating operations and improving performance. When properly conducted, the AAR is a method of sustaining successes and improving areas that are not meeting the standard. Although it is a "military tool," it is useful in any operation or action taken. The key to the success of the AAR is honesty and openness. A truly successful AAR has all the participants in the organization participating. This is sometimes a high hurdle to clear. Egos have no place in leadership and definitely have no place in an AAR. The goal of the AAR is to look at what went right and what went wrong — *not who did something wrong or missed an assignment.* The only time you would look at who missed an assignment would be if the reason for missing the assignment was a lack of training. In this case, the missed assignment is a symptom of the cause — the lack of training or possibly the lack of effective leadership.

The first official documented After Action Review was written by a Medal of Honor recipient from the American Civil War, Joshua Chamberlain. His AAR was published by the Maine Historical Society as a narrative of the actions taken by the soldiers of the 20th Maine Regiment at the Battle of Gettysburg and was actually written four days after the battle.

The purpose of this AAR was to report to superiors what actually happened at the Battle of Little Round Top compared to the orders/mission given to the 20th

Maine Regiment. The purpose of an AAR in today's business should be the same. Today's AAR follows these steps:

1. What was the actual mission/operation that we were supposed to accomplish? What was the goal of the promotion or new product fielding?

2. What actually happened? Here it is imperative that everyone has a chance to speak freely and honestly about what really happened. The leadership of the organization has to set the stage for this to be successful. In an Army AAR at the U.S. Army National Training Center, each and every soldier in the operation is given a chance to speak, and they are encouraged to speak frankly about what happened in the operation. The key to a successful AAR is the honest exchange of information. If you do not want to hear what went wrong or not strictly according to the marketing plan, promotion plan, new facility plan, then do not do an AAR — simply get your "yes" men together to make you feel good.

3. What went right or according to the plan? My experience shows that even in the worst mishaps there is almost always something that went well. Look for the good. What did we do right that we need to sustain for future events/operations? How do we sustain these things? Again, it is important to let everyone speak. Sometimes in conducting AARs you will hear someone say something that is obviously an opinion and not fact; try to keep the comments to facts. What did you see? What did we do? How do we ensure that we do the right things again?

4. What did not go according to the plan or just went wrong? In a perfect world this would not be necessary. However, because there is no perfect world — at least not on this planet — we have to look at what did not go according to plan. Was the plan flawed from the beginning? Did we not do our marketing research properly? Did the product not meet customer expectations? Did the product not function as advertised or intended? Did the promotion not target the right customer base? Did we venture outside our core competencies? Again, it must be stressed that brutal honesty is important here. Employees involved in the project should not feel threatened to speak out. Managers and leaders should not be afraid to hear feedback. After all, the life of the company may depend on fixing whatever caused the operation not to meet expectations.

5. If we did not meet expectations or did not accomplish the mission, how do we fix that for the next time? Do we need to do better research? Do we need to implement better training? Did we not communicate the plan effectively? Notice that during this step we did not mention or ask who did something wrong. Why? The purpose of the AAR is to fix the process and sustain those areas that are according to plan or better than expected. The AAR is not a tool to fix blame. My experience has been that companies that are concerned with fixing blame are not concerned with fixing problems.

6. Who is responsible for ensuring that the fix is accomplished? This is not fixing the blame or an attempt to make one person a scapegoat. The purpose of this step is to ensure that someone has the rose pinned on him/her to make sure the fix is implemented. Why? We all know that we only do well what the boss checks. So, if no one is checking to make sure the changes are implemented, how will we know that the changes will be successful? Identifying problems is easy. Just listen to the talk around the coffeepot on Monday morning after a "Football Sunday." Every Monday morning quarterback knows exactly what went wrong with his/her favorite team. Does that mean that the "experts" know how to fix the problems and make the teams better? They sure think they do; but if they could fix those problems, they would probably be in that business and not talking about it around the coffeepot. This step is important. Without this step there is no reason to conduct an AAR.[12]

Matters of Vital Importance

Throughout *The Art of War*, a common thread is the need to focus on what is of vital importance. For your company, the matters of vital importance are tied to your core competencies and to taking care of your people. How many areas do you measure or track in your organization? Are they all vital, or are they measured because they are easy to measure and "we've always measured that"? Are the matters that consume your time vital to the success of the organization, or just urgent in nature because they are the firefight of the day? Are all the reports that you prepare vital? I once worked for a boss who had stacks of reports all over his large office. As his Chief Operating Officer, I was not convinced that they were all important to the operations of the organization, but in his defense he knew every bit of data on every single report.

In another organization where I worked, there were more than a hundred "vital" weekly and monthly supply chain reports. When I took over the supply chain operations for this company as the Chief Logistics Officer, I did not believe that the boss had the time to read that many reports, knowing his scope of responsibilities. I chose the four or five reports that I thought he needed to know and be informed about, and waited for his feedback and request for the other approximately ninety-five reports. Over the next two years, he asked for only one other report. To this day, I am not sure what was done with the other "vital" reports — my guess is they were not read and were recycled. How many companies are like this and creating way too many "vital" reports that consume precious man-hours and are not perceived as vital to the health of the organization by the bosses who receive them — and in some cases do not understand them and are afraid to admit that they need some coaching of their own?

Clear Communications

Another thread woven throughout *The Art of War* is the need for clear, concise communications. How many times have you received guidance that made no sense but to your boss it was perfectly clear what she/he wanted done. In Chapter 3, Sun Tzu wrote that "The Grand Duke said, 'one who is confused in purpose cannot respond to the enemy.'" Although I am fully aware that the customer is not the enemy (but based on some of my recent experiences in the service industry, this may be a false belief in some companies), in today's business operations an employee who does not fully understand his/her purpose will not be able to respond to the customer. And in today's business climate, unless you operate in a monopoly, if you do not respond properly to customers, they will go elsewhere. In the seminal work entitled *The 7 Habits of Highly Effective People,* Stephen Covey elaborates on this concept with his principle of "seeking first to understand and then be understood."[13] Both Sun Tzu and Covey are saying that you must communicate clearly to be successful or effective. Communications, to be understood, must be clearly stated, clearly articulated, and clearly understood by those people who have to implement the instructions. Just because you understand what you are saying does not mean that it is clearly understood. Throughout history there are examples of companies and organizations that met with failure because communications were not fully understood.

Prior to what is now known to every U.S. student as the Battle of the Little Big Horn, General George Custer sent a note to his support units to "Bring Pax Quik." Historians believe his note was to quickly bring forward the pack mules with additional ammunition and supplies plus additional soldiers to the location where the eventual massacre occurred. Would this have made a difference in the course of history or the outcome of the battle? I guess we will never know because the receiver of the note did not know what Custer wanted and therefore did nothing.

In 1961, President John F. Kennedy made the proclaimed, "We endeavor to go to the moon and return safely in this decade." This intent, although short, was clearly stated, clearly articulated, and clearly understood by the employees of the National Aeronautical and Space Administration, McDonnell Douglas, and Rockwell International. On July 16, 1969, Neil Armstrong did indeed step foot on the surface of the moon, followed by Buzz Aldrin, and then along with Mike Collins returned safely to Earth.

Clear, concise, and understood communications make the difference between success and failure in operations and business. Just because it makes sense to you and you understand what the message is, does not mean that the receiver understands the message. There are more than a few companies that think they impress the employees by using big words and long messages. Some companies never know why they do not get big contracts — in more than a few cases it is because of the flowery language used rather than just answering the requirements of the Request For Proposal. You certainly do not want employees or potential clients wondering,

"What did he/she say or mean by that?" "Keep it Simple Stupid" is not just a cliché; it is good guidance for all communications.

Leadership

Another thread woven throughout *The Art of War* is leadership. Sun Tzu clearly understood the concept of influencing people by providing purpose, direction, and motivation, and the connection between strong leadership and success in operations. He called it generalship although experience over the past twenty-nine years shows that all generals are not necessarily leaders. In Chapter 3, Sun Tzu tells us that leaders are the assistants of the company. "When their assistance is complete" the company is strong and when their "assistance is defective" the company is weak. Examples such as Enron and the mess that company created and the downfall of The Home Depot after Bernie Marcus and Arthur Blank turned over the reins to Robert Nardelli 2500 years after Sun Tzu show that this wisdom is as true in the twenty-first century AD as it was in 500 BC.

Sun Tzu also tells us, "Therefore, leaders who understand strategy preside over the destiny of the people, and determine the stability or instability of the organization." In today's business, the same is true; if a leader understands the strategy of the company, he/she can and will lead the company to new levels of excellence. The Home Depot is also an example of a leader not knowing the strategy of the company. When the new CEO took over The Home Depot in 2001, he did not understand the focus of the company and the employees, and lost the focus of the customers. The result was a loss of market share, a loss of customers to the competition (Lowe's), and a decline in share price. One particular Home Depot store went through sixteen department heads in a period of a year for a store with only five departments. Some analysts have called this "toxic leadership."

Sun Tzu provides a good guideline for leaders in business today that is just as relevant in the twenty-first century as it was in Sun Tzu's time. The writings of Sun Tzu provide a historical foundation for the study of leadership. The specifics cited in this chapter show the reader that although *The Art of War* was written more than 2500 years ago, there is a great deal of relevance for leaders in this short, ancient Chinese book on warfare. *The Art of War* is mandatory reading in many Asia-Pacific Rim business schools and should be required reading for business schools in the United States and Europe as well.

Notes

1. Kaufman, Stephen, *The Art of War — The Definitive Interpretation*, Charles Tuttle and Company, Boston, 1996, p. ix.
2. The War College is the senior professional development school for officers with more than 20 years of military experience. The goal of the War College program is to prepare senior leaders for planning and operational assignments. The curriculum focuses on studies of military theory and thought and then uses a seminar approach to discuss the application of the theories to modern-day operations.

3. Taken from the Lionel Giles 1910 translation.
4. http://www.sonshi.com is a Web site dedicated to the writings of Sun Tzu and the interpretations of *The Art of War.*
5. This action in today's society is much like the practice in professional sports when a head coach/manager is fired because the team is not winning, or in some cases winning but not enough to satisfy the owner. After all, the owner cannot admit that he/she signed the wrong players who cannot function as a team so therefore it has to be the head coach or manager's fault. In major defense contracts, I refer to this same methodology as the Frankenstein methodology. Large companies compete for a government contract and then leave the majority of the workers in place and put a new head on the project by replacing the leadership team. What you sometimes get is a new leadership team or head coach who can indeed provide a new purpose, direction, and motivation and sometimes what you get is what you always got with the same workforce.
6. http://www.sonshi.com/sun2.html.
7. A SWOT Analysis is a detailed look at the Strengths, Weaknesses, Opportunities, and Threats. In this case, it is a personal SWOT Analysis and not a corporate analysis.
8. The Gospel of Mark, Chapter 9, Verse 23, *Good News for Modern Man.*
9. Kaufman, Stephen, *The Art of War — The Definitive Interpretation,* Charles Tuttle and Company, Boston, 1996, p. 18.
10. Matthew 7, Verse 11, *The Bible, New American Standard Version.*
11. The "colors" of an army could be the flags or guidons of the unit, much like today's military units, or could be the uniforms of the unit.
12. Appendix 1 details the steps and preparation in conducting an After Action Review.
13. Covey, Stephen R. 1989. *The 7 Habits of Highly Effective People,* FranklinCovey, Salt Lake City, UT.

Chapter 3

Six Sigma Leadership

There are books on the shelves of most major bookstores on leading Six Sigma and leading Six Sigma programs and projects. *Six Sigma Leadership* has nothing to do with leading Six Sigma projects. It does have everything to do with leading people to new levels of excellence in any organization.

What Is Six Sigma Leadership?

How does the philosophy of 3.2 defects per one million opportunities fit into a discussion of leadership? Are we talking about a return to the 1980s philosophy of "zero defects" for leaders? Absolutely not! That would assume that there is the perfect leader out there somewhere. Six Sigma leadership is a new twist on leader development that applies the Define-Measure-Analyze-Improve-Control (DMAIC) quality methodology to leadership. This chapter helps set the foundation for the study and benchmarking of leadership in Section II.

Six Sigma Leadership is about applying the principles of Six Sigma to leading the people of any organization. It is important to understand that leaders do not lead organizations; they lead the people within the organization. All too often this point is misunderstood at every level of some organizations.

To discuss Six Sigma Leadership, we use as a baseline the definitions of leadership from Chapter 1. As discussed in Chapter 1, all too often the terms "leadership" and "management" are confused by people who really should know better. This may be due to an assumption that leadership and management are the same thing or could very well be due to the fact that "we have always done that," thus making it "right." If this is the case, it is a good example of doing things wrong for so long that wrong looks right.

Six Sigma as a management tool is the most common use of the Motorola-designed Six Sigma methodology to remove variability and improve quality. The goal of Six Sigma Leadership is very close — to remove variability in leadership and improve leadership qualities while training new leaders in an organization.

To discuss Six Sigma Leadership, it is necessary to set a foundation of what Six Sigma really is and how it came about. Six Sigma was developed by Motorola in 1986. It started as a continuous quality improvement technique. According to Tom McCarty in an article in *EuropeanCEO*, Six Sigma "has evolved into a fully integrated management system to execute business strategy."[1] Six Sigma uses a framework of Define, Measure, Analyze, Improve, and Control. For this analysis of leadership we will use the same five areas.

Define, Measure, Analyze, Improve, Control (DMAIC)

Define

What is *leadership*? We defined leadership in Chapter 1 as "is *influencing* people — by providing *purpose, direction, and motivation* — while operating to accomplish the mission and improving the organization."

"Great leaders inspire their teams to believe so deeply in their mission that they become immersed in what they're doing…. Average leaders inspire their subordinates to do just enough to get by, just enough to get raises or keep their jobs…. Bad leaders destroy their followers' sense of commitment."[2] What type of motivation do you provide your employees? Are they content to adopt the mantra of one of my former roommates? His famous quote was always: "Minimum effort, mediocre results — there is no excuse for excellence." While I am not sure that he really believed that, he did say it so often that it made me wonder. Or, are they motivated to "become immersed in what they're doing?"

The Define phase of DMAIC includes looking at what is important. In looking at Six Sigma Leadership, this includes not only defining leadership, but looking at the question: "What do my employees expect from me?" In DMAIC, the questions include what is it that the customer wants? In Six Sigma Leadership, the customers for leadership are the employees and the company that the leader works for. To determine what "we can do to meet the customers' needs," as leaders we must know where we are in the process. We must ensure that our actions, our guidance, and our vision for the company and the future are couched in terms that the employees understand. It is imperative in leadership to make sure that we are speaking to our employees through their favorite radio station, WIIFM (What's In It For Me?).

Measure

Can you really measure leadership? Absolutely! Some organizations, including the U.S. Army, have implemented the use of 360-degree leader assessments. This is not a new technique; the U.S. Army Ranger School and at one time the Reserve Officers' Training Course have used peer ratings to supplement the assessments of their superiors. The 360-degree assessment takes it one step further by adding subordinate assessments in addition to the ratings from superiors and peers. This is one way of measuring leadership. Chapter 15 looks at some of the critical metrics for leadership.

Think about leaders who you have worked for, worked with, or observed during the course of your career. The success of subordinates is a measure of leadership. Look at some of the great National Football League coaches and then look at how many of their assistant coaches have risen to the ranks of head coaches. This is because in addition to coaching the players, the good head coaches are also coaching, teaching, and mentoring their assistant coaches. As a coach of the All Army Powerlifting Team and later as the coach of the National Champion Women's Armed Forces Powerlifting Team, I gained more pleasure from watching my athletes win the national championships than I did from my own personal national and international championships.

Analyze

To analyze your leadership and move toward Six Sigma Leadership, you must establish the metrics from the preceding section. You cannot jump from Define to Analyze. Just like the Six Sigma process improvement methodology, you have to move sequentially from one step to the next. Once you have established your leadership baseline with the world-class metrics above, you have to do a gap analysis between world-class leadership and where you are today. The 360-degree assessment is one way of validating the gap. In the Analyze phase, you determine the areas that you need to focus on to improve your personal leadership and to improve the leadership of your organization.

Part of the Analyze phase is to look at the performance of each individual on your team. All too often, experience shows that the only time supervisors sit down with individuals is once a year at annual performance appraisal time. This is the wrong time to analyze performance. As you shift into the Analyze phase of Six Sigma Leadership, it is time to sit down with every employee and assess their strengths and weaknesses. How can you evaluate an employee against a standard of performance if you do not set the standard at the beginning of the appraisal period and then provide routinely scheduled azimuth checks so the employee knows where he/she stands? Only when you establish each individual's baseline can you move to the Improve phase of Six Sigma Leadership.

Improve

Based on your gap analysis or individual assessment, it is time to develop a professional development plan to improve those areas you identified in the Analyze phase. How do you develop your subordinates into leaders? The first step is to sit down with each person individually and discuss his/her strengths and weaknesses and establish a training development program for him/her with goals and programs. Are you starting to get the point that it is important in Six Sigma Leadership to sit down with your employees on a regular basis to evaluate performance, give direction, and coach them?

The next step is to allow your subordinates to make decisions and put them into leadership positions. The Supply Chain Leadership Institute has a track record of assisting companies in designing leadership professional development programs. Part of the Improve phase is to establish new programs to bring your organization closer to world class with the above metrics and attributes.

Inherent in the Improve phase are focused training programs. Looking at the FORTUNE© list of best companies to work for each year, there is a direct correlation between the amount and quality of training that the top 100 companies offer their employees and the satisfaction the employees have in their jobs and their companies.

Control

During this phase of the Six Sigma Leadership program, the goal is to implement the professional development plans for every leader in the organization and for the future leaders to develop them and prepare them for positions of greater responsibility in the organization. During this phase, a monthly face-to-face session is necessary with each of your subordinate leaders. The U.S. Army finally made it mandatory for all leaders to sit down at least once a quarter with every employee who they rated or "Senior Rated" on annual performance appraisals. Why did the U.S. Army make this mandatory? Because if they did not, the leaders in many cases would never talk to their employees unless they messed up something. Did this solve the problem? No! Contrary to popular opinion, not every person can or should be a leader. The true leaders that I have encountered did conduct face-to-face meetings with their subordinates. Unfortunately, too many just initialed the form without ever really talking to the subordinate. Does the face-to-face meeting have to take place in the office? No. Some of the best counseling sessions that I had with subordinates were conducted while walking across the maintenance yards or walking through the distribution centers. Doing these sessions out of the office occasionally allowed me to use real-life examples and visuals as training aids.

When you move to the Control phase, it is time to move to mentoring and coaching. Please do not mistake this leadership phase for the traditional controlling that forces people to do something that they do not really want to do. That form of control is contradictory to the quality leadership that we are trying to achieve.

In Six Sigma Leadership, the Control phase is where you implement the programs that you developed from the Measure phase and the Improve phase. During this phase, it is imperative to institutionalize your development, coaching, and mentoring. During your monthly sessions, it is time to assess progress and develop corrective actions to get the employee back on track if necessary. If these monthly sessions are properly conducted, when it comes time for promotion or annual performance appraisal time, there should not be any surprises.

Coaching is just as important in business — the ultimate team sport — as it is in athletics. The principles for Six Sigma Leadership are based on my experiences in athletics as a national and international champion and as a coach of teams from Little League to National Championship Powerlifting Teams. Coaching and Six Sigma Leadership both focus on people. As an athlete, my goal was to always win when I stepped on the lifting platform. To get to a national-level caliber meant focusing on the small aspects of the three major lifts in powerlifting.

Six Sigma Leadership involves focusing on the little aspects of each individual's performance. Just as one size does not fit all when coaching a National Championship Team, the same is true for leading a group of individuals comprising a team in business. Although there is only one standard, there are as many ways of achieving excellence in individual performance as there are individuals on your team. This is where the weekly or monthly face-to-face performance counseling sessions tie in.

As a battalion commander and later as a brigade commander, I made a point of sitting down with my next-level commanders and supervisors on a weekly and monthly basis — with some it was on a daily basis. The purpose of these weekly and monthly sessions was to set the goals for the next period while reviewing the performance of the previous period. As a coach I did this after every practice with my athletes to make sure we focused on those little aspects and techniques that could make the difference between them being average athletes or champions.

Six Sigma has produced phenomenal improvements in quality, customer service, and supply chain operations. Using this adaptation of Six Sigma for leadership will produce phenomenal improvements in your organizations that will bleed over into other operations and activities. Using Six Sigma Leadership in your organization will eliminate the poor leadership at all levels that all of us have seen in organizations. Some of these include surprises at performance appraisal time, not really knowing what the boss wants — "I'll know it when I see it," and leaders who come right out and say, "I am more concerned about my career than I am about yours."

The bottom line is that a focus on leadership will produce the following equation:

$P3 = P!$
People + Passion + Pride = Profits.

When you focus on the left-hand side of the equation, the right-hand side will improve. However, if you focus purely on the right-hand side of the equation, you may fail on both sides of the equation. One company told me that it did not have time to focus on people because it was so focused on profits. That company was in bankruptcy. I could not resist stating the obvious — "Maybe you are focusing on the wrong thing. Because you are not making profits, perhaps a shift in focus is necessary. Focus on the people and the profits will come because the employees will take pride in the company and will work to make it better."

Implementing Six Sigma Leadership will enable you to achieve operational excellence in your operations. Remember that regardless of what business you think you are in, *you are in the people business and that requires leadership.*

Section II looks at the values and attributes of world-class leadership and provides the framework for benchmarking your leadership as well as providing the framework for developing leaders in your organization — this is the legacy that you leave your company.

Notes

1. Tom McCarty, *EuropeanCEO*, September–October 2004, Leadership Edition, "Six Sigma® At Motorola."
2. Smith, Dean and Bell, Gerald D., *The Carolina Way*, The Penguin Group, New York, NY, 2003, p. 33.

LEADERSHIP©: THE ATTRIBUTES OF LEADERSHIP

Leadership is the art of getting someone else to do something you want done because he wants to do it.

—General Dwight D. Eisenhower in a speech to his soldiers prior to the Normandy Invasion.

This section looks at the attributes and values of world class leaders in all walks of life. A leader can be anyone who is in a position of responsibility and has the capability to influence others. Regardless of the level that a person leads in an organization, the attributes discussed in this section are applicable. As a leader grows in responsibility or moves up the corporate ladder, the depth of the impact of his/her actions influence subordinates may increase.

In the US Army's Field Manual for Leadership there is a lengthy discussion of the concept of BE-KNOW-DO.* The BE aspect includes the ingrained virtues or attributes of world class leaders as discussed in the building of the House of Leadership.

The KNOW aspect includes the technical and tactical aspects of the industry. What is important to understand and will be discussed in more detail during the chapters on the attributes of leadership is that employees really do not care how much you know about the job until they know how much you care about the people doing the job.

The DO aspect of leadership is how leaders respond to situations and pressure. When I was in Kuwait preparing the distribution operations for Operation Iraqi

---

* US Army Field Manual 6-22, October 2006, p. 1-1.

Freedom, it was interesting to watch how "leaders" at all levels reacted to the stress and pressure of getting ready for combat. One particular high ranking "leader" came completely unglued one night in the Dining Facility when a SCUD missile alarm went off. Other leaders reacted much better to the stress and pressure. As leaders develop and benchmark themselves against the attributes of leadership, they will respond better to the pressure of leading others and accomplishing the missions while improving the organization. The attributes that form the House of Leadership could be considered the core competencies of leaders. They are not a check the block list of competencies or attributes.

The United States Air Force manual on leadership states, "The abilities of a leader, which are derived from innate capabilities and built from experience, education, and training, can be improved upon through deliberate development."* The House of Leadership and the attributes that form the House of Leadership provide you with the foundations for deliberate development. This deliberate development of leaders in your organization will take individual capabilities and assist you in developing world class leadership in your organization.

The House of Leadership provides a framework for developing your own leadership style, benchmarking your progress, and developing leaders in your organization. The House of Leadership will also provide you a framework for establishing the BE-KNOW-DO philosophy in your organization.

Applying the Attributes and Values of LEADERSHIP© to your organization will also enable you to prevent the onset of Motivational Dysfunction in your organization or cure Motivational Dysfunction if it already exists when you get to a new organization.

---

* US Air Force Manual, p. vi.

Chapter 4

L1
Loyalty

> Plato originally said that only a man who is just can be loyal, and that loyalty is a condition of genuine philosophy.[1]

The popular online dictionary Wikipedia defines *loyalty* as faithfulness or devotion to a person or cause.[1] Using this definition there are multiple forms of loyalty: loyalty to your country; loyalty to self, as in "to thine own self be true"; loyalty to your family, probably the oldest form of loyalty; loyalty to friends; loyalty to a company; loyalty to an employee or employer; and loyalty to your customers.

Loyalty — Up and Down the Chain

Loyalty is a critical value for leaders at all levels. Loyalty is a two-way street — a leader must be loyal up and down the chain of responsibility. All too often, leaders expect loyalty from their subordinates without showing loyalty to their subordinates or to their bosses.

How often have you heard someone in charge say, "This is not the way I would have done this but the boss said to do it this way." This is setting up the mission for failure and ensuring that the speaker is not blamed for the failure. The employees will develop an attitude of negativity and the operation/mission/product launch will fail. When it fails, the boss who made these comments will be the first to say, "I told you this would not work."

Loyalty as a virtue and leadership value would dictate that in this situation, if you do not believe a course of action will work, explain it to the boss and make sure he/she understands why the course of action will not work and how to make it work. At this point, if the boss still insists on the course of action and directs you to "make it work," you are obligated to explain the course of action to your employees and work to make it successful. This is demonstrating loyalty up the chain.

Can you be successful as a leader without loyalty up and down the chain? History provides us with examples of leaders who were not loyal up the chain. General Douglas MacArthur comes to mind. His run-in with President Harry Truman made headlines during the Korean War. In his memoirs, General Omar Bradley did not demonstrate loyalty to his superior in World War II, General Eisenhower, or his subordinates, especially General George Patton, Jr.

However, to be successful in business and still be able to look yourself in the face every morning without flinching, you have to demonstrate loyalty to your employees, your employer, and your customers. These three represent the legs of the three-legged stool of loyalty. As with any other three-legged stool, if one of the legs is shorter than the others, the stool is not balanced or level. The same is true for your leadership — if one of the legs of loyalty is shorter than the others, you will have trouble modeling leadership for your employees.

Loyalty and Employee Retention

How do you attract, develop, and retain loyal employees? The job must be fun. In fact, when the job becomes fun, it is no longer just a job. As a commander of a supply and distribution management battalion and then as the commander/CEO of a much larger supply chain organization, I told all my employees that I wanted them to look forward to work on Monday mornings. I also told them that I fully understood that by the end of the day they would probably be ready to go home, which was fine as long as the next morning they were excited about coming back to work. During both assignments to the U.S. Army's National Training Center in the Mojave Desert, I knew that a large number of my employees were driving an hour each way to work and in the summer would be working in extreme desert conditions. It was important to make sure that they were personally satisfied with both their work and the work environment, and were rewarded when the established standards were exceeded for quality support. To do this, we worked very hard and at least once a month we all got together in an informal setting to relax and get to know each other better on a personal level. What this did was build team morale while building team cohesiveness. The result was that every standard that the U.S. Army established for supply chain and maintenance support was surpassed almost every single month, and the organization set new levels of excellence for other organizations to try to emulate. It also resulted in exceeding all standards for employee retention — not because it was the only place to work, but because the team building produced loyal employees.

Another key to developing loyal employees is that they must not view their job as a dead end. They must have hope that there is a chance for promotion. Granted, there are some employees who are content at their current level and do not want a promotion. You have to deal with these employees on an individual basis. I once had a mechanic in Hawaii who had no desire to be promoted but the U.S. Army had an "up or out" policy that stated that at certain points in a career, a soldier either had to be promoted or separate from the U.S. Army. This mechanic was content to take engines apart and put them back together in much better condition than they started in. In most companies, this would be acceptable. In the U.S. Army at the time, it was not and this talented mechanic chose to separate rather than get promoted to a position that would prevent him from doing what he enjoyed most in life.

Another easy-to-implement method of attracting and retaining loyal employees is to adopt a program of "catching them doing something right." Way too many companies have what I call the "discipline Nazi" running around trying to catch someone doing something wrong and then writing them up for it. Obviously, it is important to enforce standards in a company through discipline but I have found it more effective and a lot more fun to catch someone doing something right and recognizing them for it. It may be a simple pat on the back or a reward of time off or a dinner on the town with his/her spouse or significant other.

One major company in the Midwest retains loyal employees by allowing them to do legitimate volunteer work on company time and recognizing the volunteers for the work that they do and the goodwill that it creates for the company. This is another low-cost method of developing and retaining loyal employees.

Although most employees will tell you that money is not the only motivator in staying at a company, you will also find out that not paying an employee what they are worth will lead to employee turnover. One such company in Southern California has in excess of 50 percent turnover a year. When questioning them about it, it was revealed that they paid their warehouse workers $10 an hour. They paid for their forklift training and paid for their OHSA certification on the forklifts. They also paid a bonus at the six-month mark for staying with the company. However, after the bonus was paid, employees were leaving for another distribution company a few blocks down the road. Further research revealed that this new company was paying almost twice as much for a certified forklift operator. This was not a matter of employee disloyalty but rather a matter of not paying the employee what the market value of a trained and certified forklift operator was worth.

Loyalty and Training

Providing training is a great way to foster employee loyalty. The key with training is to invest in the right employees and make sure that they have some commitment to the company when the training is complete. Here are two key examples of these principles. First, make sure the deserving employee gets the training: how many employees get to go to valuable training because you can do without that employee the easiest? I once sent an employee to a five-month school, but not because he was the most deserving of the training; in fact, he was the least deserving but I could afford to do without him and, in fact, his absence for the five months dramatically improved the efficiency of the operation and the morale of his fellow employees. I do not recommend this as a course of action. The second example is linked to the concept of ensuring that the employees stay around after the training. In addition to the example of the distribution company in Southern California, a company in Virginia paid for four employees in its IT section to become Microsoft Systems Certified. As soon as the employees received their certifications, they all took jobs at another company. I thought this was an anomaly; after all, a large company would not spend good money to train and certify Microsoft Systems Engineers. However, I recently observed a major defense contractor that sent an employee to a two-week training program and paid for his Systems Security Certification, only to have the employee take another job with a competitor company just a little over a month after becoming certified. Two points do make a line and in this case indicate that the first incident was not an isolated incident. As important as training is to attracting and retaining quality employees and as important as certifications are to employee morale and esteem, there must be a commitment from the employee and companies must ensure that only loyal and committed employees are sent to career-enhancing training programs.

Developing Loyalty

So, how do you develop loyalty in your company? Loyalty must be earned. It comes from developing a bond of trust with your employees. Your employees then form a bond of trust with your customers and everybody wins. To earn the trust and loyalty of your employees, you have to show loyalty to them and to your superiors in the company. You have to model loyalty to your employees. Your employees will then model this same loyalty to your customers and the result is customers for life. A satisfied loyal employee and a satisfied loyal customer are the best advertisements your company can have.

Can a loyal, enthusiastic employee really make a difference? Here is an example. In the Leavenworth, Kansas, Home Depot, a department head in the flooring, wall, and décor department would go to other stores to pick up items for customers. These customers would only do business with this associate and this led to the associate being ranked first in the state of Kansas for sales. Customers would ask for this associate by name and, if not available, would come back later. How did the store reward this employee? They did so by canceling her vacation that had already been approved and paid for. And the manager who made the decision had an assistant manager tell her instead of doing it himself. The associate quit after six years with the company — a company that for two years she drove seventy-five miles each way to continue working for the company while her husband was stationed at a remote Army base in the Mojave Desert. How did the customers react? They took their business elsewhere and two years later they still ask other associates when she is coming back. The store continues to treat its employees the same.[2] Again, the customers were not there out of loyalty; they were there only because it was convenient.

Your employees are your company to the customers. If you have a product delivered from your store or warehouse to a customer, the delivery team is the company to the family receiving the product. If your employee is not happy, your customers will know it. Do you really want an employee who embodies the attitude that one lot employee at a major home improvement company explained to me? Obviously, customer satisfaction was not his concern. He told me, "The first and the fifteenth I get paid whether I bust my butt or not." Do you think this leads to customer loyalty? No way. The same company tried to implement a program of more part-time employees under a new CEO program to save money. The concept was to hire more part-time employees, thus reducing benefits costs. The problem took several months to surface — the part-time employees did not have benefits, nor did they have employee stock options and ownership. This simple cost-saving program cost market share, cost a loss of stock value, and cost the company in other intangible ways such as the loss of formerly loyal employees who were serviced by less than loyal part-time employees without a stake in the company. This program to save money took almost a year to reverse once the real results surfaced.

Loyalty from a leader to his/her employees leads to trust between the employee and the leader. Employees who trust their leaders become loyal employees and will be motivated to work toward a common goal.

Communication is key to developing loyal employees. Remember the example of General Grant — make sure your employees always know what you want, when you want it, and why you want it. Another way of looking at the communications piece is to remember Emerson's five wise men — who, what, when, where, and why. When communicating visions, mission statements, or any other corporate communications, always ask, "Who else needs to know?"

The *Stanford Encyclopedia of Philosophy* states that "Loyalty is usually seen as a virtue … families expect it, organizations demand it."[3] Demanding loyalty may be a policy of companies but to develop loyalty in employees and customers, you have to demonstrate loyalty to the employees and customers. Loyalty may be expected but it cannot be demanded. If loyalty becomes demanded, does it become blind loyalty as was demonstrated in the late 1970s in the Jonestown incident where the entire colony drank cyanide-laced Kool-Aid? Can you be loyal sometimes? No. Would you expect your spouse to be loyal and faithful only sometimes? Probably not. Partial loyalty is not loyalty at all. Is it possible to be loyal to a fault? This is possible. Employees or members of an organization become so loyal that they defend the organization even when they know the actions of the organization are wrong. History is full of examples of people blindly following a cause or a "leader" in a gang or crowd who they would normally not follow if they were by themselves. The actions of the Nazi Party come to mind. The members of the Party may not have shared the same beliefs of Hitler and his cronies but because they remained loyal to the Party (most probably out of fear), they became guilty by association. Many leaders of the Confederate States of America during the American Civil War fell into the category of being loyal to a cause that may have, in fact, been contrary to their personal views. Many of these leaders placed their loyalty to their state above their loyalty to their country. Robert E. Lee and Thomas J. Jackson are examples of this. If you believe that the American Civil War was all about slavery, these two leaders did not believe in that but did believe in their home state of Virginia.

Loyalty and Customer Retention

Keeping customer loyalty, like employee loyalty, is also a benefit to the company financially. Like employee retention, keeping a loyal customer is cheaper than developing a new loyal customer. In 1998, the Shoney's in Prince George, Virginia, burned down. This left two choices for the local lunch crowd if they wanted a salad bar for lunch: The Western Sizzlin' and The Pizza Hut. The Western Sizzlin' offered the larger salad bar and also had a hot bar and dessert bar for those not quite as health conscious as the salad bar crowd. Conversely, Pizza Hut offered a lunchtime pizza buffet. The majority of the business from Shoney's went to the Western Sizzlin'. Western Sizzlin' had customers but they were not loyal customers; they were customers of convenience. They could have become loyal customers if Western Sizzlin' had tried to cultivate the new customers. Unfortunately, they assumed that they had a monopoly without Shoney's and did not improve their offerings; in fact, the quality of their food declined over the next few months. When Shoney's reopened, the loyal Shoney's customers returned to Shoney's and today the Western Sizzlin' that had been in place for almost 20 years is closed.

In an article online, Aaron Green stated that "You can't have loyal customers without loyal employees."[4] This ties employee attitude and loyalty to customer satisfaction and loyalty. Sam Walton used to say that it takes about two weeks for the attitude that you show your employees to show up in the attitude your employees show your customers. So there is a tie between leadership, employee loyalty and attitude, and customer loyalty and attitude toward your company and your brands. Green went on to state that "The concept of loyalty applies to employees as well. Just because someone has worked for your organization for twenty years does not necessarily mean he/she is loyal." Sometimes it is easier to stay at one job that is not satisfactory than to try to find a new job. This leads to motivational dysfunction, which embodies itself in the employees' attitude shown to your customers.

Does loyalty to employees foster employee retention? Employee retention is a key indicator of leadership effectiveness. It is much easier and definitely much cheaper to retain an employee than it is to recruit and train a new employee. Is there a difference between a loyal employee and employee longevity? There could be but in some cases, disloyal employees may stick around for a long time because they have no other place to work.

Does customer retention equate to customer loyalty? Maybe, but in some cases there may not be a choice. When I lived in Virginia, a bad ice storm left most of the area that I lived in without power for several days. The result was that we all had to purchase kerosene heaters to keep our houses warm. One particular large home improvement big box store had no competition in the area and made the customers feel like the store was doing them a favor by letting them shop there. The customers remained not out of loyalty, but out of need. When a competitor moved into the area a few months later, most of the customers moved to the new store.

As a company, it is the things that you do for your customers that you do not have to do that will make a difference between developing loyal customers and just customers of convenience, as demonstrated by the home improvement company above. Here is an example of a company going out of its way for a new customer. *Fancy Gloves* based out of North Carolina has a very good Web site. I placed an order for formal gloves for my youngest daughter for her homecoming dance. I did not check the order for accuracy before hitting the Submit button. The red gloves that I thought I had ordered were actually ordered as purple gloves — not a good match for a black dress with a red sash. When the gloves arrived and I discovered my mistake, I called the company. Because it was a weekend, I got the voicemail of the owner and left a message describing my situation with the dance only a week away. Early Monday morning, I received a call on my cell phone as well as my home phone. The owner explained that the length that I had ordered was now out of stock in red but she would substitute a more expensive glove without additional charge on the verbal agreement that I would send back the purple gloves. Not only did she not charge for the more expensive gloves, but she also express mailed the gloves so that my daughter would have the right color for the dance. This is going above and beyond the expectations of the customer.

Sometimes larger companies believe that customers will continue to use them even when service does not meet expectations. In 2006 I used FTD.com for flowers for my wife for our anniversary and for her birthday. The flowers for her birthday arrived without a problem. Three weeks later for our anniversary, the florist company tried to deliver the flowers to the wrong address and then called to say that they could deliver them on the following Monday — only a few days late. It does not take a genius to know that flowers a few days late for an anniversary are not a good idea. However, in an attempt to give this same florist another chance, again they told me that the flowers were not deliverable because the address did not exist. This was after I called and spoke to a customer representative and explained what had happened the previous year and that I wanted to confirm the address. They still changed the city and missed the delivery for my wife's birthday. You can believe another company was used for our twenty-fifth anniversary flowers.

There is a connection between customer loyalty and employee loyalty, and it is a leadership responsibility to develop loyalty in the organization. Customer and employee loyalty are not sometimes things and must be earned every day. Loyalty is a cornerstone of successful organizations, regardless of their size.

Summary

There is a direct link between the loyalty that you show your employees, the loyalty that you show your employer, the loyalty that you show your customer, and the profitability of the company.

When your employees are willing to go out of their way to take care of a customer, it is a good sign that a leader has gone out of his/her way to take care of that employee. When employees stay at a job because of the loyalty shown by the leadership rather than just for the money, the employees have developed a sense of loyalty. And when employees are concerned because they feel like they let down their leadership, you have developed loyal employees. When you model loyalty, your employees will work hard to emulate that behavior and will reward you with loyalty.

Customer and employee loyalty are impacted by the ethics, integrity, and honesty demonstrated by you and your organization or the perception of ethics, integrity, and honesty you model for them.

Notes

1. http://www.en.wikipedia.com/wiki/loyalty, accessed March 1, 2008.
2. I know this story to be factual because the associate/department head in this story is my wife Kay. This same store went through an average of four department heads for each department over the period of about eighteen months — all after a new manager came into the store. The assistant manager in this story told his employees that he was not paid to wait on customers and do physical labor, but he was paid to delegate.
3. http://plato.stanford.edu/entries/loyalty/
4. Accessed February 22, 2007.

Chapter 4 Questions

1. What sense of loyalty do your employees have? Are they willing to work to get the job done, or do they jump ship at the first good offer from another company?
2. What sense of loyalty are you modeling for your employees? Do they see a leader who is not loyal to his or her employer and emulate that lack of loyalty to you?
3. Is there really a link between employer loyalty and employee retention?

Chapter 5

E + 1
Ethics and Honesty

> Simply put, ethics involved learning what is right or wrong, and then doing the right thing — but 'the right thing' is not nearly as straightforward as conveyed in a great deal of business ethics literature.[1]

There is no right way to do a wrong thing. *Ethics* and *honesty* are two of the most important of all the skills and attributes of leadership. These two attributes are related and cannot be compromised by leaders if they want to be effective. In addition, ethics, integrity, and honesty by leaders contribute to employee loyalty, which as we have discussed leads to customer loyalty. Ethics, integrity, and honesty are key to success: you cannot stretch the truth, or live on the edge of ethics and integrity. You have to set the example for your company and your employees. People are watching you to see how you react/respond to situations. Some authors have written about situational ethics — there is no such thing; there are ethics or there are not. Few things in business are simply black and white — ethics, integrity, and honesty are clearly black and white; there are no half truths or partial ethics.

Every company should be concerned about ethics. According to Carter McNamara, "far too many resources about business ethics end up being designed primarily for philosophers, academics, and social critics. As a result, leaders and managers struggle to really be able to make use of the resources at all."[1] The goal of this chapter is to provide useful information for leaders of any size organization that can be applied in developing ethical leaders and ethical companies.

> Ethical behavior is an investment in the long term.[2]

Ethics in Supply Chains

Why are ethics important to supply chains? Let's start by defining what ethics are and then we can look at the importance of ethics in supply chains.

What are *ethics*? This word is used almost to excess in today's business environment. To study ethics, we need to define them. The Dictionary.com Web site defines ethics as "the rules of conduct recognized in respect to a particular class of human actions or a particular group, culture, etc."[3] The *Merriam-Webster Dictionary* defines ethics as "the discipline of dealing with what is good and bad; the principles of conduct governing an individual or group."[4]

How do you measure ethics if you are benchmarking yourself or your employees? Ethics, like honesty and integrity, is a go/no-go measurement — either you have ethics or you do not; there is no in-between. I am convinced that we are all born with an ethical compass that points "true North." I am sure I am not the only person who has accidentally or intentionally demagnetized a compass. If you have never done that, please take my word for what happens. When a compass is demagnetized, it will never point to "true North." Therefore, if you are depending on the compass for directions, you will probably never get to where you wanted to go. Unfortunately, many people today have somehow demagnetized their ethical compass. Just like the compass that I demagnetized, their compasses just spin and point in any direction they want to. When this happens to individuals, they start making unethical decisions. This then becomes a downward spiral and, like the Enron scandal, they take the company and its employees down with them. There is only one measure for a leader's ethics. Ethics are not something that can be turned on and off like a light switch. They must be turned on and stay on.

Ethics is a measure of your ability to determine right from wrong. This is something most of us learn as children and hopefully remember when we move into the real world. Ethics is not a measure of your morality, although the two are related. Morality is a symptom of ethical behavior, or a lack thereof. This makes ethics easy to model.

The lack of ethics seems to be recognized quickly by the press. I am not sure if it is because there are so many examples almost daily of individuals, companies, and governmental officials who exhibit a lack of ethics or because the press is always looking for the bad in everything. It appears that sometimes the press even gets caught up in the lack of ethics and integrity that they enjoy reporting on. Two good examples have surfaced in the past several years.

During the 2004 U.S. Presidential Campaign, Dan Rather became so intent on reporting anything bad on the candidates that he allowed himself to compromise his ethics and integrity in reporting "confirmed" information about President George Bush's military record. When this was proven false, Dan Rather lost all credibility as a reporter despite his several decades of reporting.

As this chapter was being written, the Department of Homeland Security and the Federal Emergency Management Agency (FEMA) displayed the utmost in violated ethics and integrity. FEMA, still reeling from the debacle of its efforts in response to Hurricane Katrina in 2005,[5] was seemingly in control of the efforts to respond to the 2007 California wildfires. These wildfires resulted in the largest evacuation in the history of California as over a half a million individuals were forced to evacuate the rapidly spreading destruction. FEMA decided to hold a press conference to discuss the evacuation and the successes of FEMA in managing the evacuations and the responses in fighting the fires. Unbeknownst to people watching the press conference and the reporters who were forced to teleconference into the press conference due to short notice was that the "reporters" asking the questions were FEMA employees. The result of the press conference was originally thought to be very favorable. Less than twenty-four hours later, the truth was leaked to the press and, once again, FEMA and the Department of Homeland Security lost credibility because of a lack of ethics and integrity.

Although ethics gets a lot of attention in business today, it is not a new subject. Plato and Aristotle spoke and wrote about ethics and honesty thousands of years ago. Some truths are timeless. The truths of ethics, integrity, and honesty fall into this category. Are there differences between business ethics, contracting ethics, and personal ethics? No, there is one standard for ethics.

The Ethics Resource Center conducted a large-scale study of ethics in business in 1994. The purpose of this study was to determine the impacts of corporate ethics programs on employees and employee behavior. Another goal of the study was to "identify the major ethical issues facing employees today."[6]

The results of this study showed that the number of companies with corporate ethics programs continues to increase while employees doubt the existence of rewards for ethical behavior and found that employees were skeptical of the ethics of their co-workers. Because of this mistrust of the company to reward ethical behavior, employees were also leery of the ability of their company to handle ethics problems. It could be that in these companies, the company and the employees did things wrong for so long that wrong started looking right. This is not a good situation in which to be.

How many times have you seen unethical behavior or actions at work and did not report it? This could be as simple as using company assets for personal gain or "fudging" on the hours accountability such as recoding a full day's work when only a partial day was worked — "no one will know." You know and that is enough. Employees are watching their leaders and managers for examples of how ethics decisions are made. Your example will speak much louder than the words of the ethics program. If a leader is not setting the example, the ethics program is not worth the paper on which it is written.

Aristotle said that "we study ethics in order to improve our lives, and therefore its principle concern is the nature of human well-being."[7] As leaders we are concerned about the well-being of our company and our employees. Therefore, ethical behavior is a must.

Here is an example of a corporate ethics program gone bad. This company was trying to promote ethics by having an "Ethics Week." The goal of the Ethics Week

```
┌─────────────────────────┐
│ Ethics Week Challenge   │
│  August 25–29, 2008     │
└─────────────────────────┴──────────────────────────────────┐
  **How it works:**

  I receive an award if I am the Business Conduct Officer who
  "meets and greets" the most number of employees (you) during
  Ethics Week.

  You receive an award if your are one of the first 500
  Employees who respond indicating that you were greeted
  by their Business Conduct Officer (me). Remember to mention
  my name.

  **How to participate:**
  Provide your name, ID#, and mailing information along with my name.


  Your Name _____
  ID # _____

  Mailing Info: _____
  _____
  Business Conduct Officer Name: _____
  _____


  Fax or e-mail this info to:
  Fax: (xxx) xxx-xxxx        E-mail: EthicsLink@xxxxxxxxx.com
```

Figure 5.1   Ethics Week contest.

was to get the ethics officers out to meet the employees and let them know who the employees could turn to when they (the employees) saw or heard of unethical behavior. The card in Figure 5.1 was handed out to supervisors with the name of the ethics officer on the other side. The supervisors were told to hand out the cards to their employees.

Notice that the card states "I receive an award if I am the Business Conduct Officer who 'meets and greets' the most employees." First off, the ethics officer passed these out to the supervisors to give to the employees — not exactly ethical to give ethics officers an award for "meeting and greeting" if they did not actually meet or greet the employees. The second part of this card that bothers me is the employee getting an award for being one of the first 500 to report that they were, in fact, greeted when they were not really greeted by their ethics officer.

If you are going to have an ethics program or an ethics promotion in your supply chain for your employees, please at least make sure that it does not appear to be unethical in its approach as this program appeared to be.

In one of his concerts, Jerry Clower, the late, famous country comedian, gave the audience a good bit of advice in between some of his funny stories. The advice that he gave on one particular night was about decision making and ethics. Basically, what he told the crowd was that "if you have to think about it, you are fixin' to mess up." The same is true in ethics today. If you have to ask, "Is this ethical?," you probably already know that the action or decision is not ethical and in some cases are just looking for someone to say, "Go ahead and do it."

Following are some examples of ethics codes and regulations from opposite sides of the United States.

The state of New Jersey provides this example of statutes on ethics for government employees with comments from the author on why these are important in the commercial world as well. The New Jersey State Statutes contains the following paragraph on ethics in describing the Ethics Commission:

9:61-2.3 Plain Language Ethics Guide

The Commission shall prepare a plain language ethics guide which provides a clear and concise summary of the laws, regulations, codes, orders, procedures, advisory opinions and rulings concerning ethical standards applicable to State officials. The guide shall be prepared to promote ethical day-to-day decision making, to give general advice regarding conduct and situations, to provide easy reference to sources, and to explain the role, activities and jurisdiction of the Commission.[8]

Why is it important to have ethics rules in "plain language?" It is important because the state of New Jersey wants to make sure that there is no gray area in the ethics guidelines and statutes for government officials.

The plain language guide contains the following Principles of Ethical Conduct[9]:

- "You may not use your position to secure a job, contract, governmental approval or special benefit for yourself, a friend, or family member." Why is this important? Obviously sometime in the past, this was a problem and New Jersey did not want there to be an appearance of undue influence. Does this happen in business today? Absolutely! That is why more and more companies are establishing their own codes of ethics.
- "Your paycheck is your only permitted compensation.... You may not 'moonlight' without the approval of your agency." The bottom line here is that employees are being paid to do a job, and anything related to that job should already be covered in their salary. The part about moonlighting is to ensure that the work the employee or official performs for the state does not influence work outside the office. The U.S. Army used to prohibit moonlighting by soldiers without the approval of their unit. The reason for this was twofold: (1) to ensure that the soldier was not doing a job that would bring discredit to the U.S. Army, and (2) to ensure that the unit knew where the soldier was in case of an emergency or unit deployment. Companies

sometimes become concerned when employees have a second job. This concern is based on conflicts of interest. Which job is more important when making trade-off decisions? Can you serve two masters? Does this mean that working a second job is unethical? By no means. Many people find that a second job is necessary to make ends meet in today's society. The key is to be sure that the jobs do not compete and cause a conflict of interest.

■ "You may not accept any gift of more than nominal value." Most companies are now putting a dollar value on the "nominal value" rather than leaving it up to the employee to determine if the gift is of nominal value or not. The best policy is to prohibit accepting gifts of any kind from suppliers or contractors to prevent the illusion of impropriety.

■ "You may not be 'wined and dined' by people with whom your agency conducts business." This is a good policy for all companies. Allowing employees to be "wined and dined" by potential suppliers can easily give the illusion of unethical behavior. One particular large organization conducts an annual conference for senior military officers and historically includes drinks and hospitality suites to hopefully gain an inside advantage for future contracts and acquisitions. Is this ethical? It has been blessed by ethics lawyers but still gives the impression of unethical behavior when those same officers eventually take jobs with those companies. One particular problem with this form of activity is that it may be completely innocent but may give the impression that something is not completely ethical. I have a very close friend and former workout partner who became a government contractor with a company that later did business at an installation where I had control of all logistics and supply chain activities and employed large numbers of government contractors. Although his company did not do supply chain contracts, to prohibit the appearance of impropriety, when we would meet for lunch on his visits to my location, he would insist on separate checks for the meals. His concern was to make sure no one could possibly believe that there was anything unethical occurring. On the other hand, I found it amazing how many "friends" and "associates" who I had not heard from in several years all of a sudden wanted to stop in for a chat when a $200 million contract for my organization was released for proposals. And you have to understand that to "stop in" where I was at the time meant driving more than two hours from the closest airport or, even if you were in the area, you were still forty-five minutes away at the closest freeway exit. At this location no one just "happened to be in the area" or just "stopped by." So, I referred them to the contracting office for "a chat."

■ Prior business relationships. "You may not have any involvement on official matters that involve any private sector individual or entity that employed you or did business with you during the one year prior to the commencement of your State employment." Although it may be very innocent and the prior employer is the only company that produces that product, there is still the illusion of impropriety. It is similar to giving large, noncompetitive contracts to your former employer, as happened when Halliburton/Kellogg, Brown, and Root received extremely large support contracts for Operation Iraqi Freedom. There are only a few companies that have the capability to provide the types of support needed for soldiers, airmen, and marines in the magnitude that was needed for Operation Iraqi Freedom, but the fact that the Vice President of the United States was the Chief Executive Officer of Halliburton prior to becoming the Vice President certainly gave the pundits the impression that the contracting process was not ethical. Some companies use non-compete clauses and non-disclosure clauses to prevent employees from taking business to another company or taking clients to another company to prevent unethical behavior claims.

■ "When in doubt, ask." This is always a good policy. Even when you are sure that what you are doing is completely ethical, if a member of your team is in doubt, ask your ethics advisor.

An area that is definitely unethical is the popular "unnamed source" or "high-ranking official" who leaks information to the public. In some cases this may be done to allow information to reach the public. However, in most cases, an unnamed source is releasing information that is closely held and would not otherwise reach the public. Releasing information in this way to the public or to a reporter to broadcast is unethical and can in some cases cause irreparable damage to an operation or activity.

The state of Hawaii addresses this very problem in its Standards of Conduct: "No employee shall disclose information which by law or practice is not available to the public and which the employee acquires in the course of the employee's official duties, or use the information for the employee's personal gain or for the benefit of anyone."[10] Perhaps some U.S. Government employees and appointees should be held to the same standard as the state of Hawaii employees and legislators. I am sure other states have the same standards for their employees and government officials. The Hawaii Standards of Conduct are used here as an example.

The state of Hawaii's preamble to the State's Standards of Conduct states:

> The purpose of this chapter is to (1) prescribe a code of ethics for elected officers and public employees of the State as mandated by the people of the State of Hawaii in the Hawaii Constitution, Article XIV; (2) educate the citizenry with respect to ethics in government; and (3) establish and ethics commission…. This chapter shall be liberally construed to promote high standards of ethical conduct in state government.[11]

Every company should have the same goal for its standards of conduct and ethics management programs — to promote the high standards of ethical conduct.

The State of Hawaii Ethics Commission members are not allowed to hold any other state office and are not paid by the state of Hawaii for their services. Both of these practices help ensure the integrity and ethical appearance of the commission. In contrast, most corporations have members of their boards of directors as members of their ethics committees. One large FORTUNE 500© company only addressed ethics in light of the Securities and Exchange Commission guidelines and regulations. Although this is very important, there is more to ethics in business than just the SEC. This is only part of a company's ethics. Ethics covers all aspects of a company and is a daily responsibility of every single employee. However, it is the leaders who have to model what ethical behavior looks like for the employees to emulate.

The Institute for Supply Management (formerly the National Association of Purchasing Management) takes ethics and integrity in their profession very seriously. Take a look at their Code of Ethics[12]:

> Loyalty to your organization; Justice to those with whom you deal; Faith in your profession.
>
> Avoid the intent and appearance of unethical or compromising practice in relationships, actions, and communications.
>
> Avoid any personal business or professional activity that would create a conflict between personal interests and the interests of the employer.
>
> Promote positive supplier relationships through courtesy and impartiality.
>
> Know and obey the letter and spirit of the laws applicable to supply management.
>
> Develop and maintain professional competence — ties to self development.
>
> Preamble —
>
> A distinguishing characteristic of a profession is that its practitioners combine ethical standards with the performance of technical skills. In fact, "professional" is described in *Webster's New Collegiate Dictionary* as "characterized by or conforming to the technical or ethical standards of a profession." … In order to achieve stature as a profession, those in supply management must establish and subscribe to a set of ethical standards to guide individual and group decisions and actions.

Another key to ethics in an organization is how the company deals with unethical behavior. Is it simply swept under the rug, or does the organization face it head-on. My experience from an early job that included vacuuming floors at the Amoco Oil Credit Card Center helped me learn early that when someone sweeps something under the rug, it leaves a lump or wrinkle in the rug that someone else (in this case, me with my vacuum) has to clean up. Ethics is very similar to my job with the vacuum cleaner — if you simply sweep it under the rug, someone else will have to clean it up and then your integrity will be suspect as well. On the weekend shift at Amoco, I frequently found stuff literally swept under the rugs that my co-workers and I had to vacuum — these piles were obvious wrinkles in the rug. Unethical behavior is the same — you have to deal with it immediately. It is not another shift's responsibility to clean up after you.

Is ethics as simple as doing what is right and not doing what is wrong? Can anything be that simple? In fact, it is simply doing what is right. There is no right way to do a wrong thing. Unfortunately, for some companies the issue of ethics is only a problem if the issue gets them on the evening news — otherwise it is okay. This is analogous to sweeping the issue under the rug. Your word is your reputation to the business world; if that word is no good because of ethics problems, you are going to be out of business. The name Enron will forever be linked to unethical behavior. In the movie *Scarface*, Tony Montana stated that all he had was his word and he would not break it for anyone. Before the movie is over, he does break his word and it results in his death. Now this is the movies and obviously his business was not ethical to begin with, but the same is true for all of us. All we really have is our word and if we engage in unethical behavior, we no longer have that. The true lesson from this example is that if you engage in unethical behavior and break your word, you may not die physically but you will die in the business world.

What about ethics in goal setting and performance bonuses? Although we will look at goal setting as part of the discussion on determination and dedication, it is important here to look at the link between ethics and goal setting. One major U.S. Department of Defense contractor sends out a letter on ethics and integrity every quarter with an annual letter to employees on how well the company performed on its ethics goals for the year.

Within a few weeks of the latest letter from the company President and Chief Executive Officer, a strange ethical dilemma surfaced in one of the company's major divisions. The CEO of the division was rumored to be on the way out because it was announced by the division CEO that none of his subordinate officers made their goals for profitability for the fiscal year. Within just a few weeks of the rumors of his imminent departure, *Forbes Magazine* reported that the division CEO actually received a $2 million bonus for making his profitability goals from the parent company. Research and questions into the matter revealed that, apparently, this division CEO set artificially high goals for his subordinates, which meant that none of them received a performance bonus while his goals were indeed met and a significant bonus ensued. Ethics is not a situational attribute. Actions such as this cause employees to lose all respect for their bosses.

Ethics in supply chains continues to garner significant attention. In 1986, a major supplier to the U.S. National Aeronautic and Space Administration (NASA) was made aware of a significant defect in the O-rings that were used as a seal on the Space Shuttle's external fuel tanks. The supplier and NASA apparently did not think that the defect was serious enough even though the O-ring was known to weaken in extremely cold temperatures. This defect was basically "swept under the rug" and was not considered a big deal. At least it was not a big deal until the morning of January 28, 1986. On that chilly Florida morning, the whole world watched in horror as the Space Shuttle Challenger exploded less than two minutes after clearing the launch tower at the Kennedy Space Center. Did a supply chain ethics violation lead to this disaster? One could definitely make a case that if the supplier had displayed supply chain leadership and supply chain ethics, the seven astronauts would still be alive today. Would a stronger sense of ethics in the Morton-Thiokol supply chain have prevented this disaster? Will a stronger sense of ethics and modeling ethics for your employees prevent a supply chain disaster in your company? Ethics usually only becomes an issue when ethics are violated. Make it a priority in your company; do not wait until your company is on the front page of *USA Today* before ethics is important in your company.

Leaders must set the example for modeling ethics. As a leader, everyone is watching you to see how you react or act in certain situations, to include ethical behavior. You cannot adopt the attitude of "Oh, nobody will know." How many of us thought that when we were growing up? And yet somehow, Mom and Dad always seemed to know. When your employees are watching your every move and every action, someone will know. I once had an employee who kept a journal on everything I did and said. He was always trying to catch me doing something that he did not perceive as being ethical. I am proud to say that he was not successful in his quest. Your behavior will serve as the model for your employees.

Even companies that pride themselves in high ethics, such as Wal-Mart, sometimes find themselves with unethical employees. In 2005, Wal-Mart let a vice chairman of the board go over misuse of over $500,000. His defense? He was using the money to keep tabs on union activities in the stores — another unethical practice. The moral here is that all companies can have ethics dilemmas and therefore need a strong ethics training program and leaders modeling ethical behavior.

Here is a simple acid test for ethical behavior from the Canadian Association Xpertise Inc.:

> If you're unsure whether or not an action is ethical, there's a simple acid test to you. Ask yourself the question, "Would I have any objection to having my actions detailed on the front page of a national newspaper?" If you can honestly answer "No" then the chances are good that you're acting ethically.

Do you do something for someone in hopes that it will provide you with a benefit? If the answer is yes, then you need to ask yourself if that is ethical. Here is an example from my past. When I was a Reserve Officers' Training Course cadet at North Carolina State University, I had a professor of military science who modeled what unethical behavior looked like. His first action (one that I later found was very common) was to provide a scholarship to the son of a high-ranking military officer over several more qualified cadets. The next year as this same lieutenant colonel was preparing for retirement, he started showing favoritism toward another cadet when he found out that the father of this cadet was a high-level appointee in the North Carolina Department of Transportation. This lieutenant colonel intimated

> **Dear Colleagues:**
>
> Great companies are defined by their reputation for ethics and integrity in every aspect of their business. By their actions, these companies demonstrate the values that serve as the foundation of their culture and attract the best customers, employees and stakeholders in their industry.

Figure 5.2   Introduction to the Sprint Nextel Code of Conduct.

that he really wanted to work for the State Department of Transportation when he retired. This favoritism was blatantly obvious to everyone in the Military Science Department at the university. As fate would have it, around the time that he was preparing to retire, the cadet's father retired also without considering the lieutenant colonel for a job. Wasted, unethical behavior was rewarded properly in this case. The interesting thing about this situation was that the two cadets that the lieutenant colonel compromised his ethics for both chose not to go into the Army.

Most companies have an ethics program or an ethics policy, and many are adopting ethics committees. One major defense and aerospace contractor's annual ethics letter ties its ethics to its corporate goals. The annual letter states that

> Ethics are the foundation for the five objectives of our corporate vision … and for the corporate vision to be the most trusted provider in our industry.… Our ethics define our "character" as individuals and as an organization. Ethical conduct is about doing what is right, being accountable for our actions and holding ourselves to the highest standards of behavior.

Do you have a code of ethics for your supply chain? Figure 5.2 provides an example of a corporate code of conduct from the Sprint Nextel code of conduct.

The Sprint code of ethics continues with a very good series of questions about what is ethical, as shown in Figure 5.3. And Figure 5.4 provides yet another section of Sprint's code of conduct.

Making Ethical Decisions…When in Doubt, Ask Yourself…
- Could it harm Sprint Nextel's reputation?
- Is it ethical and legal?
- What would my family and friends say?
- How would it look in the newspaper?
- Would I best my job on it?
- Should I check with my supervisor?
- How would my action appear to others?

Figure 5.3   Sprint Nextel Code of Ethics questions.

We Lead By Example

Supervisors are expected to exemplify the highest standards of ethical business conduct by integrating the ethics and compliance program into all aspects of their operations and by encouraging open and frank discussion of the ethical and legal implications of business

Figure 5.4   Sprint Nextel code of conduct for leaders.

What makes Figure 5.4 critical to the discussion on modeling leadership for employees is the statement that "supervisors are expected to exemplify the highest standards of ethical business conduct." Why? Because employees are watching the leadership to see how it responds and acts in business situations so the employees will know how to respond in similar situations. This is modeling the behavior for employees to emulate.

Ethics and Contracting/Procurement/Acquisition

Vice President Spiro T. Agnew came under investigation for allegedly receiving payoffs from engineers seeking contracts when he was Baltimore County executive and during his tenure as Maryland's governor. He was also accused of taking kickbacks in the White House, leading to his resignation. There was a lot of speculation at the time that this was a diversionary tactic to take the eyes off of the other unethical behaviors that were apparently present in the Nixon Administration that eventually led to the resignation of the President of the United States.

The U.S. Government takes ethics in procurement and contracting very seriously. The U.S. Office of Federal Procurement Policy Act is part of United States Law. As such, it places restrictions on disclosing and obtaining contractor bid or proposal information or source selection information for government contracts. Why? The U.S. Government is concerned about leaking insider information and giving a company an unfair advantage in contracting and bidding for projects. The basis for this law is that the U.S. Government is concerned about ethics in contracting. I realize that ethics in governmental actions is considered by some as an anomaly or an oxymoron, but there are laws to protect the integrity and ethics in contracting. Concern for ethics in contracting goes a little farther — the Federal Acquisition Regulation states, "an employee shall not, other than as provided by law, knowingly disclose contractor bid or proposal information or source selection information during the three-year period after the end of the assignment of such employee." This prevents insider information influencing other bids or contracts of a similar nature. If the U.S. Government is concerned about procurement and contracting ethics, shouldn't your company be just as concerned?

On the flip side of the coin, the Federal Procurement Policy Act also states that "A person shall not, other than as provided by law, knowingly obtain contractor bid or proposal information or source selection information before the award of a Federal Agency procurement contract to which the information relates." Again, the U.S. Government is concerned about the ethics, integrity, and honesty on both sides of the contracting equation.

All too often, contracting personnel will try to gain an advantage over the competition. Here is an example of what I considered a violation of contracting ethics. Prior to returning to the U.S. Army's National Training Center, I was contacted by an old friend who had retired a few years earlier from government service. I had worked with him and his wife during my previous assignment at the Training Center. I naively assumed that because we were moving back into the area, that he was just being friendly. After a meal at his new house and a few "friendly" visits over the course of the next few months, I discovered his true agenda.

Unbeknownst to me, this "friend" had signed on with one of the potential contractors for the $250 million contract. As a consultant for this company, his plan was to use our "friendship" to attempt to gather information on the contract. When I discovered his ulterior motive, I suggested that future visits until the new contract was let were probably not a good idea because it might give the illusion of insider information. Amazingly, after his company did not win the contract, I never heard from him again. His company's vice president, whom I also had known for several years, also stopped calling when his company did not get the contract — he had gone as far as promising employment after retirement, which I now realize was a promise tied to his company winning the contract re-bid.

Hopefully, you will never be in the same situation; but if you find a "friend" from the past suddenly showing up when a large contract for your company is announced, be leery. During this particular solicitation and bid process, several "old friends" came out of the woodwork calling to see how I was doing.

The Federal Procurement Policy Act also establishes criminal penalties for misconduct:

> Whoever engages in conduct constituting a violation of this section for the purpose of either —
>
> (A) Exchanging the information covered by such section for anything of value, or
> (B) Obtaining or giving anyone a competitive advantage in the award of a Federal agency procurement contract, shall be imprisoned for not more that 5 years or fined, or both.

The reason for these criminal penalties is to prevent even the illusion of wrongdoing or unethical behavior. What policies do you have in place to prevent the illusion of wrongdoing or unethical behavior?

What if the mores of a country differ from yours for handling business? In some countries, accepting a gift is commonplace. I once worked with a company in a foreign country that even offered one of my employees the favors of "ladies of the evening" in exchange for work. How do you handle such a situation? What do you do when your ethics conflict with the mores of the host country or host company? Remember that your employees are watching what you do and what behavior you model for them. Dr. Robert Handfield wrote about this a few years ago[13]:

> I was present at a holiday dinner during the 1970's, also attended by my company's Director of Manufacturing and Director of Engineering, their wives, and hosted by the owner of an international business partner company also accompanied by his wife. We had just initiated a significant joint venture with this company, and the dinner was a type of "celebration" for the new partnership.
>
> I was the highest-ranking official representing the company in the room. At the beginning of dinner, the host walked into the room with a flourish, and produced a beautiful set of earrings, offering them to the wife of the Director of Engineering. She was clearly delighted. Next, he produced a beautiful amulet, which he placed around the neck of the wife of the Director of Manufacturing. She also thanked him profusely, and accepted the gift. Next, he walked over and opened a box, which contained an exquisite diamond necklace, that must have cost tens of thousands of dollars, and placed it around my wife's neck. She politely thanked him, held her hand to her throat and stated that she could not accept the gift. He insisted, saying that it was "his birthday", and for that reason, she must accept it as a gift to him. I intervened, and noted with some firmness that we definitely could not accept the gift. The dinner was taking place in a large public restaurant. The discourse had attracted some attention for nearby diners. Later at dinner, he tried again to present the necklace, and did so again at dessert. Each time, I politely declined. Finally, as we were leaving, he stated that if we were not going to accept the necklace, then he would leave it on the floor, which he did. We left shortly thereafter.
>
> Was this inappropriate behavior on my part? We clearly had insulted a business partner – yet did I act with integrity? Was it the right decision? To answer this question, consider the following context…
>
> I reported to our CEO for 10 years, having been promoted to VP at age 34. During this time I had accountability for purchasing, traffic and logistics for the entire organization. Depending upon the time period within the 10 years my additional responsibilities included: R&D, formation of a new division with P&L, management of a self contained process control computer manufacturing division, energy management, real estate management and management of the corporate office building including a relocation of 300 people to a new office. Numerous other issues of ethics, internal policy interpretation, social responsibility

and inter-company turf were encountered. The single best complimentary letter that I ever received came in a year ago from my former CEO as I began a new business venture: The note included this line: "Your reputation for integrity will serve you well."

During my years as VP and CPO, my President and Chairman were each quite active in the civic and business community. My Chairman was one of the founders of the newly formed National Minority Purchasing Council, the fore-runner of today's NMSDC, and was a social responsibility advocate by word and deed, and served several other local city boards in our corporate home city. My President and CEO had an armful of outside activities. He had served as a Campaign Finance Chairman for a U.S. Congressman, chaired the State Industrial Development Commission for the Governor, was president of the local Chamber of Commerce, served on several corporate boards, was chairman of the industry trade association, and was often on the phone with high profile politicians who would ask his opinion or seek input and support on economic and political issues.

The point of all of this foundation building is this. Both men were in constant communication and meetings with others who were seeking to do business with our company. In one of the wisest meetings I have ever attended, the three of us met early in my tenure. The gist of the meeting was this.

1. We meet people who ask how they can sell something to our company. These are our friends, our business associates, our fellow club or church members and our political allies. We need to be nice to them.
2. You are in charge of buying things and determining who we should buy from.
3. We will refer anyone who asks for an opportunity to do business with the company to you.
4. You will ignore any implications of referral or recommendation that you hear from these people and make the best business decision for the company.
5. If our own mother should come in to sell pencils and her price/terms are too high, politely "throw her out."
6. Our mutual job is to build relationships and to do the best job we can for the company.

Much of what I see about corporate ethics is policy, implementation, judgment and accountabilty driven. The above scenario adds something to the mix. There should properly be an internal understanding and partnership that establishes an ethics policy but also manages the policy in a manner that, in the borrowed words of one line of Rotary International's 4-way Test, "Builds good will and promotes fellowship."

One can say "No" in many alternate ways. Ethics is about avoiding the appearance of conflict of interest, not just the conflict itself.

Returning to the business dinner and attempted gift of jewelry, the decision to refuse the gift was a good one. As CPO, my personal reputation would have been compromised by appearing to accept such a gift in a public place even if it were discretely returned at a later time. The acceptance of gifts by other corporate personnel created an uncomfortable precedent; yet it was my job to do the right thing. Finally, the reader should know that the business deal later fell apart based, in part, on poor business practice by the international "partner." In this situation, one had only a few seconds to make a decison that must be "lived with," explained to others and support continued business relationships. My wife and I had thoroughly discussed issues of family and business ethics; she was an equal partner in making the correct decision with conviction and grace.

Unethical behavior can have some severe consequences. In these examples, the offenders were people who should have or did know that what they were doing was not ethical or legal. Following are some examples of unethical behavior from actions taken to the U.S. judicial system:

- *United States v. Randall Cunningham.* In this case, former Congressman Cunningham accepted bribes from contractors and other interested parties "in exchange for taking official action to influence the appropriation of funds and the execution of Government contracts" that benefited the parties paying the bribes that represented large defense contractors. The former congressman was sentenced to one hundred months in prison and three years of probation for accepting an inflated price for his California home, a paid-off mortgage on another home, a downpayment on a new home, while also accepting a yacht, a Rolls-Royce, and a paid party for his daughter's graduation.
- *United States v. Karla R. Kronstein, United States v. Michael G. Kronstein,* and *United States v. Kenneth N. Harvey.* In 1998, while serving as the Chief of the Acquisition Logistics and Field Support Branch for the U.S. Army Intelligence and Security Command, Harvey recommended a large contract be awarded as a sole-source contract[14] in exchange for an offer of employment with the Kronstein's company and approximately \$40,000 in payments to Harvey's wife. The courts sentenced Mr. Kronstein to seventy months in prison and Mr. Harvey to seventy-two months in prison plus restitution. Mrs. Kronstein waived her spousal rights and testified as part of a plea bargain.
- *United States v. Lester Crawford.* Crawford served in multiple senior positions with the U.S. Food and Drug Administration. Apparently, he "forgot" to disclose that he held significant stock and stock options through

his wife in PepsiCo, Sysco, Kimberly-Clark, and Embrex — all regulated organizations under the regulatory responsibility of the FDA. As a result of his position as the Director of the FDA, he participated in an obesity study group that impacted companies in which he and his wife had a financial interest. As a result of his conflicts of interest, lack of ethics, and false statements, Crawford resigned and received three years of probation.

The purpose of showing these sample cases involving contracting ethics, or a lack thereof, is to show that the U.S. Government does take ethics in contracting very seriously. How seriously does your company take ethics in contracting? What ethical behavior are you modeling for your employees? How does your company handle ethics violations? Do they take action or simply sweep it under the rug? Does the position of the person committing the unethical act influence what actions are taken?

Title 5, Volume 3, of the United States Code, entitled "Administrative Personnel," contains Chapter XLVII — "Federal Trade Commission." Under this is Part 5701 — "Supplemental Standards of Ethical Conduct for Employees of the Federal Trade Commission." These standards are established to prevent the illusion of unfair or unethical practices by members of the Federal Trade Commission and establishes guidelines for getting approval prior to any outside employment. This review by the Agency Ethics Official is to ensure that members of the commission do not take outside employment that would compromise their ability to serve on the commission.

Ethics and Transportation Contracts

Is there a problem in this area? Apparently a recent problem surfaced in Hawaii with the transporting of supporters and other interested fans to the 2008 Sugar Bowl in New Orleans. The Hawaii Ethics Commission is looking into the transportation costs for the supporters, government officials, and faculty/staff. The ethics rules state that personnel can only have their airfare paid to such an event if and only if they are on official state duties. The *Honolulu Star-Bulletin* and the *Honolulu Advertiser* had to file Freedom of Information Act paperwork to get the names. The first list had forty-five names blacked out. After a lengthy back-and-forth process, the University of Hawaii agreed to release the complete list only after talking with the university's lawyers. Ethics violations can take seemingly innocent turns — or at least "innocent" in someone's eyes. In this case, someone should have realized that a free flight from Honolulu to New Orleans for a football game was not completely ethical. Even politicians are starting to question some of the "perks" of the office when they include free meals, free flights, and gifts. But then again, as noted here, some politicians and political appointees are still having problems recognizing ethical behavior.

A General Accountability Office (GAO) audit of a defense contractor revealed "that by relying on employees to monitor their own behavior, the company increased the risk of non-compliance, due to either employees' willful misconduct or failure to understand complex ethics rules." The review identified lack of management controls as a weakness in the company's ethics program.

According to the Associated Press on June 13, 2008, in an article entitled "Army Contracts Budget Soars, But Watchdogs Are Held Steady," the U.S. Army's contract costs have almost tripled since 2001 while the number of auditors has stayed the same. The Army is trying to catch double-billing, accidental and intended, bribes, and contract kickbacks. One former senior Department of Defense official was quoted as saying, "What we really need to eliminate abuse is people doing it right the first time." This is what we all need as we model contracting and procurement ethics. If we model ethical behavior at all levels of our supply chain, we will indeed have people doing it right the first time.

Just a few days earlier, *The New York Times* reported that a U.S. Army colonel had pled guilty to accepting bribes, a vacation to Thailand, and a travel trailer in exchange for releasing insider information on warehousing contracts in Iraq. Do you have someone in your organization who is releasing confidential information, intentionally or accidentally, to your supply chain partners that impacts your procurements and contracts? Have you recently walked through the process or audited the process to find out?

According to the Federal Procurement Policy Act, one of the most important purposes of ethics rules and laws is to assist employees in avoiding conflicts of interest or the illusion of conflicts of interest. If the U.S. government takes ethics in contracting and acquisitions so seriously, should you as a leader of supply chain employees be just as concerned about ethics in your supply chain? Do you have policies in place to provide guidelines to your employees to prevent them from being placed in conflicting situations?

The Institute for Supply Management is so concerned about ethics in procurement and supply chain practices that in 2005 it published "Principles and Standards of Ethical Supply Management Conduct." The purpose of these principles is to promote loyalty to the organization; to promote a standard of justice in all supply chain dealings; and to foster faith in profession of supply chain practitioners. The goal of the Institute's principles and standards is to:

- Avoid the intent and appearance of unethical or compromising practice in relationships, actions, and communications.
- Avoid any personal business or professional activity that would create a conflict between personal interests and the interests of the employer.
- Promote positive supplier relationships through courtesy and impartiality.
- Know and obey the letter and spirit of the laws applicable to supply management.
- Develop and maintain professional competence — this ties to a later chapter on self-development.

The Preamble to the Institute for Supply Management's "Principles and Standards of Ethical Supply Management Conduct" states:

A distinguishing characteristic of a profession is that its practitioners combine ethical standards with the performance of technical skills. In fact, "professional" is described in *Webster's New Collegiate Dictionary* as "characterized by or conforming to the technical or ethical standards of a profession." ... In order to achieve stature as a profession, those in supply management must establish and subscribe to a set of ethical standards to guide individual and group decisions and actions.[12]

Ethics policies are in place in many organizations. We have looked at the standards from Sprint and the preamble to the standards for the Institute for Supply Management, but how valuable are ethics policies? The goal of such policies is to promote ethical standards for benchmarking in your organization and prevent the perception of unethical behavior. Perceived ethics problems are just as damaging as real ethics violations.

Some companies use surveys to measure the success of their ethics programs. However, ethical surveys and ethics hotlines are only good if the employees see actions come from them. The same is true for reporting ethics violations; what action is taken will determine if future reports are made. Actions — not just words!

How do you track the ethics of your employees? One major defense contractor requires all its employees to take annual ethics refresher training (the training is always the same, so apparently it is not important enough to update it annually with new situations and cases). After the annual refresher training, this company then has the employees fill out a questionnaire. Included in the questionnaire are questions such as: "Did you in the past year release confidential information from Company X to competitors?" and "During the past year did you accept any gifts from companies that wanted to do business with Company X?" When I see such questionnaires, my first thought is that if the employee did violate standards of ethics, why does the company believe that the employee will now be ethical enough to admit to it?

Someone is always watching. Is your loyalty to the company part of ethics? Do you take supplies home? Do you order specific expensive supplies because you like that brand better than the standard brand? This is why I have used my own pens at work since I was a lieutenant — I did not want anyone to think I was taking the pens home. I did not want the illusion of impropriety. I had heard too many horror stories while I was in the Officers' Basic Course.

How do you represent the company to the customer? You are not just an agent, but also a representative of the company. Avoid activities that would compromise, or create the perception of compromising, the best interests of the employer.

Ethics and Expense Reporting

Where does the requirement to be prudent stop and the trust of your employees begin? How careful are you in looking at expense reports for your employees? Do you and your company automatically assume that an employee is unethical and dishonest, or do you put some trust in your employees until they prove otherwise?

When I was a young company commander, I had the members of the Presidential Campaign Communications Support Team attached to my company at Fort Gordon.[16] This team was a hand-picked team of highly qualified non-commissioned officers that established and supported all the communications needs for the Presidential candidates and their support teams and Secret Service detachments. Needless to say, this was a very high honor for these soldiers to be part of such a team.

After one very long trip, one of the soldiers submitted his travel voucher to the U.S. Army for reimbursement of his traveling expenses. When the travel voucher reached the Finance and Accounting Office, it was referred to the U.S. Army Criminal Investigation Division for potential fraud. It seems that the soldier had received such good service from the housekeeping department at one hotel that he left the traditional tip for the housekeepers and claimed this on his voucher. When I reviewed the voucher, it seemed legitimate to me (but remember I was just a young captain at the time). The amount averaged to about $2.00 a day for the three-week stay.

After a very lengthy and costly investigation that included sending an investigator to interview the housekeeping staff (now that is close to waste, fraud, and abuse), the investigation was closed and no action taken because the housekeepers said that they had indeed received a nice tip from the soldier. The tip worked out to about $450.00 on a voucher for almost $10,000 in travel expenses. The investigation took several months and thousands of dollars because someone was not willing to rely on the ethics and integrity of the soldier. Unless your supply chain employees have proven otherwise, expect ethical behavior; and if they are proven otherwise, get rid of them; there are enough things to worry about in supply chains without having to worry about a few employees skimming products and money from the company.

I later experienced the same wrath from the accounting office. I was asked to come to Florida at the invitation of the U.S. Army Recruiting Command to give powerlifting demonstrations and a short talk about athletics and the military. I averaged seven demonstrations and talks a day for a week at a time. Needless to say, putting on seven demonstrations[16] and maintaining the energy to do so required more than the normal amount of energy. So, I usually ate five to six smaller meals during the day to maintain the energy and strength to make the demonstrations impressive for the high school students.

When I returned from one such trip, I was called to the legal office to justify why my meals list showed five to six meals a day and why my breakfast meal costs exceeded the recommended levels for expenses. I had to get a statement from the hotel (which was a bit secluded) to show that the only breakfast it had was a buffet that indeed exceeded the recommended cost for breakfast. Then I had to write a sworn statement for each meal (not one sworn statement for all the meals, but one for each meal for a week) that I was in fact not trying to cheat the U.S. Army out of a few dollars. I also had to get a statement from the recruiting office that was driving me around that it did indeed have to take me to five to six meals a day.

Like the example of the soldier on the Presidential Campaign Communications Support Team, the investigation and time spent cost the U.S. Army much more than the original claim to prove my innocence. Trust your employees, point them in the ethical direction, model the behavior for them, and then let them show you every day that they are ethical, and not unethical.

On the flip side of the ethics coin, when on trips for the recruiting command, I often wonder how many corporate dollars are claimed on expense reports for questionable activities. Some of the hotels that the U.S. Army booked for me in Florida were not located in the resort areas of Central Florida — in fact, they were not even located in areas where most of us would want to stay. There were "gentlemen clubs" located close to more than a few of them and, as I went for my evening run, I often noticed "gentlemen" in coats and ties going into these clubs. The thought was always, "Here I am justifying an extra dollar or two for a meal and these guys are on business trips going to clubs of questionable reputation. I wonder if they are putting this on their expense reports."

Ethics is closely related to *honesty*. How does your honesty impact your leadership?

Honesty and Integrity

Always tell the truth, and if that is not possible, tell the truth anyway.
—John "Buck" O'Neil[17]

Honesty is defined by Wikipedia as "the human quality of communicating and acting truthfully related to truth as a value.... Quality of honesty applies to all behaviors."[18] The Boy Scout Law starts with "A Scout is Trustworthy." This is good guidance for anyone in business and for everyone regardless of their occupation or place in life. The only way you can be trustworthy is to follow the guidance of Buck O'Neil — always tell the truth.

As a leader modeling leadership for his/her employees, it is imperative to always tell your employees the truth. If you want them to follow you, you must be honest with them. How do you model and benchmark honesty? Do your employees and your boss believe what you tell them, or do they question you because of previous integrity issues? Benchmarking honesty is easy: either you have it or you don't. If you have established a record of honesty and integrity with your workforce, you have best-in-class honesty. Like integrity and ethics, there is only a yes or no answer to honesty as a value/attribute of leadership. Although the writings of Confucius recognized three levels of honesty, there really is only one level of honesty to benchmark.

Honesty is important in evaluations of performance and potential. Honesty in evaluations for employees is critical to improving their performance while improving the whole organization one employee at a time.

You are not doing your employees any favors by giving them a "walk-on-water" performance appraisal if they did not meet the agreed-upon expectations and levels of performance. And if employees are not meeting expectations, then annual performance appraisals should not be the first time they finds out that they are not meeting expectations. How many times have you received an annual performance appraisal and wondered why? On the other hand, how many times have you given an annual appraisal and had the employee act surprised? Even when you sit down with the employee on a regular basis, some employees will act surprised if you do not tell them that they "walk on water." I once had a deputy commander who, although I sat down with him once a month to talk about where he needed to be and what he needed to do to meet the standard and receive an outstanding appraisal, still acted surprised when he received a "satisfactory" appraisal. I showed him the records of our monthly sessions that he had signed in agreement and the proposed accomplishments for the year that he had agreed to and still he seemed surprised. This employee filed a grievance but was it was found to be unsubstantiated because of the record of the performance improvement sessions.

Give each employee an agreed-upon set of standards for performance and an agreed-upon list of goals and accomplishments for the appraisal period. Sit down with each of them on a regular basis for an azimuth check and let him/her know what is going well and what needs improvement. Be honest. Do not tell the employee that all is well if it is not; and if something needs to be improved, be honest and coach him/her on how to improve that area.

One home improvement warehouse had a store management team that was convinced that every employee had something wrong with them or their performance and made it a practice to "surprise" the employee with something new at every performance appraisal. This same store terminated an employee of nine years with a strong record of outstanding performance appraisals because he took a sick day when he repeatedly asked for approval to take a vacation day. Granted, calling in sick when you are not is not honest, but in this case the store management created a situation where the employee felt that was the only course of action. In fact, this same store allowed an assistant manager, who had tendered his resignation to work for a competitor, to write employee appraisals after the last day he was employed by the company. Ethics, integrity, honesty? What behavior was this store modeling for its employees?

Be honest with your employer. I once had a roommate in college who had a bad habit of lying to employers and professors. This individual went as far as printing up doctor's stationery to get him excused from classes. But the biggest example of going to extremes to cover a lie came one summer when we were performing throughout the state of North Carolina as part of the Governor's Official Bicentennial Drill Team. My roommate decided that he was too tired to go to work one night after a performance, so he called his boss at the restaurant he was working at and told him that he had broken his finger when the musket case (about the size of a standard coffin) fell on his hand. To cover this lie, he went to the drug store and bought a finger splint that he wore for several weeks. It would have been much easier to call and say that we were late getting back rather than having to perpetuate the lie. This individual went as far as having a fake diploma printed to show his parents so they would believe that he graduated from college. I am sorry to say that when employers started checking references, they found out that he was not being honest about his schooling and, unfortunately, my roommate with the great personality never landed a good job. His career of dishonesty caught up with him.

How does this happen? I caught the tail end of a movie on TV and heard only the answer to the question, "Why did you tell those lies?" The answer was simple: "I told one lie, then another, and then the truth did not matter anymore." Does this really happen to people in the real world? Absolutely! First a little "white lie," then another little lie that "did not hurt anybody," and then another — and before the person knows it, he/she does not know where the real truth stops and the lies begin.

Be honest with the employee. I received a lesson in honesty and integrity on performance appraisals when I was a young captain in the U.S. Army. I was being transferred to another post and my supervisor had not completed the appraisal prior to my departure.[19] My supervisor showed me what he was going to put in the appraisal and asked me to sign a couple of blank appraisal forms just in case there was an error on the one he showed me. (This was before the days of word processors or computers and we were still using typewriters). When I finally received the signed copy a few months later, the words on the appraisal were different from the ones I agreed to prior to my departure. In fact, during the year that I worked for this boss, I learned a lot of lessons in honesty, integrity, and ethics — most of them being what *not* to do.

I later learned that this lack of honesty and integrity was a family trait when I had his son as one of my company commanders when I was a brigade commander. The son had learned from his father by modeling his leadership and integrity on what he saw of his father's style. When I remarked that his signature looked like that of his father, the response was shocking. He told me, "I spent so much time trying to sign his name on my papers that I guess I mastered it."

You can always be an example. Make it a positive one. This young company commander learned from the model of his father. Unfortunately for this particular son, his learned lack of integrity and honesty resulted in him being asked to leave the Army before completing his commitment.

Summary

If you are modeling ethical behavior, your employees will see your actions and base their actions accordingly. On one deployment to Korea, I ran into an old friend who was leaving Korea and offered to sell me his "collection" of about fifteen bottles of liquor for a very small price. The price included a footlocker that was about the cost of what he wanted for all of his "collection." My boss heard about the "collection" from a friend to whom I had given a bottle and asked for a couple of the bottles. Because I did not drink that form of alcohol, I gave them to him. When I no longer had his favorite brand, he later came back and wanted to discipline me for having too much alcohol. Fortunately for me, by that time I had given it all away — but I did keep the footlocker. The ethics he was modeling did not match what he was saying.[20]

Your ethics, your honesty, and your integrity will be the benchmarks that your employees will look at to model their leadership styles. Make sure that you are setting the right example for them to model and against which to benchmark. You do not want to be the leader who everyone uses for the example of what *not* to do.

What happens in your supply chain when you compromise your ethics or practice dishonesty? What happens when ethics are violated? Enron is a classic example of a violation of ethics. WorldCom also comes to mind. The unethical practices of Enron impacted business across the globe as a result of the implementation of the Sarbanes–Oxley Act in response to Enron's "house of cards." Do not let a lack of ethics destroy your supply chain.

Notes

1. McNamara, Carter, Authenticity Consulting, LLC, *Complete Guide to Ethics Management: An Ethics Toolkit for Managers*, http://managementhelp.org/ethics/ethxgde.htm. Accessed October 7, 2007.
2. Axelrod, Alan. 2006. *Eisenhower on Leadership*, J.B. Wiley & Sons, Jossey-Bass, San Francisco, p. 40.
3. Dictionary.com. *Dictionary.com Unabridged (v.1.1)*. Random House, Inc., http://dictionary.reference.com/browse/ethics. Accessed October 3, 2007.
4. http://mw1.merriam-webster.com/dictionary/ethics. Accessed October 3, 2007.
5. Two years before Hurricane Katrina hit New Orleans, a comprehensive study was conducted by the U.S. Army Corps of Engineers on the impact and results of Category 4 Hurricane hitting the New Orleans area. The study is an amazing prediction of what actually happened. The study stated almost to the inch how high water would be in New Orleans after the levies failed from the storm backwash. Knowing this and armed with another study and after action report from the Indonesian Tsunami response, the Corps of Engineers, FEMA, and the citizens of New Orleans should have been prepared for what happened.
6. http://www.ethics.org/research/1994-study.asp
7. Aristotle's Ethics, http://plato.stanford.edu/entries/Aristotle-ethics/
8. New Jersey State Statutes, paragraph 19:61-2.3, http://nj.gov/ethics/statutes/rules/index.html. Accessed October 3, 2007.
9. State of New Jersey. September 2006. *Plain Language Guide to New Jersey's Executive Branch Ethics Standards*, State Ethics Commission, Trenton, NJ.
10. State of Hawaii. Constitution of the State of Hawaii. Chapter 84, Section 12, Standards of Conduct.

11. State of Hawaii, Constitution of the State of Hawaii. Chapter 84, Standards of Conduct.
12. Institute for Supply Management, 2005. *Principles and Standards of Ethical Supply Management Conduct.* Institute for Supply Management, Tempe, AZ, p. 2.
13. Handfield, Robert. 2004. *Ethics and Supply Management in a Global Environment,* http://scm.ncsu.edu/public/hot/hot040922.html. Accessed, July 21, 2008.
14. A sole source contract literally means that only one company is capable of providing the material or service requested by the contract. Sole sourcing is often used to ensure that a company gets the contract and the specifications are written in such a way that it limits the contracting pool to one particular company. Sole source contracts are mistaken for single source contracts. In a single source contract, there are multiple vendors but a company chooses to contract with only one company.
15. In the U.S. Military, every soldier must be under the administrative, financial, and legal control of a company. In this example, the Presidential Campaign Communications Support Team was an ad hoc team established to support the 1984 Presidential Campaign. These soldiers were "attached" to my company for all personnel, financial, training, and legal matters.
16. The demonstrations for the high school students included a bench press demonstration consisting of a few warm-ups and then a demonstration of 315 to 405 pounds for up to eight repetitions for each class followed by the talk on lifting, athletics, and Army opportunities.
17. Buck O'Neil was the first African-American to coach in Major League Baseball after the integration of the National League. O'Neil also managed the Kansas City Monarchs of the Negro Leagues to multiple championships. For more on Buck O'Neil, the Negro Leagues, or the Negro Leagues Baseball Museum in Kansas City, MO, go to http://www.nlbm.com.
18. http://en.wikipedia.org/wiki/honest. Accessed September 28, 2007.
19. By U.S. Army regulations, a soldier is supposed to have his/her performance appraisal completed and signed before he/she leaves a unit. He/She is also supposed to be formally counseled on the appraisal and what should be improved and/or sustained in the performance prior to leaving the unit.
20. This particular boss had to have a beer every day at 5:00 p.m. when the regulations allowed alcohol consumption. While in the field in Korea, we arranged for the supply sergeant to pick up a six-pack of beer every day on his supply and laundry run and place it on the boss' bed with his laundry. As long as he got his daily beer, it was easy to work with him.

Chapter 5 Questions

1. Do you have an ethics policy for your supply chain?
2. How does an ethics violation impact your supply chain, your customers, your employees?
3. How do you benchmark ethics in your supply chain?
4. What steps can you do to improve the modeling of ethical behavior for your employees?
5. Is shaving a few minutes here and a few minutes there ethical?

6. Look at the examples in this chapter and in the local newspaper, and ask yourself if these were acts of short-sightedness or acts of deliberate dishonesty.
7. Review the examples of unethical behavior in this chapter and discuss the difference between legal behavior and unethical behavior.
8. Just because another country has different business mores and standards for business ethics, does that allow you to conform to their way of doing business, or should you continue to travel on the ethical route?

Chapter 6

A3
Attitude, Aptitude, Accountability

If you think you can or if you think you can't, you're right.
—Henry Ford

"You've got an attitude." How many times have you heard that or said that to an employee or child? The truth is that we all have an attitude. The key to modeling leadership is to model a positive attitude. Attitudes are as contagious as the common cold. That being the case, what do you want your employees to catch — a good, positive, customer-centric attitude or an attitude of indifference? The choice is yours as a leader.

Attitude

What is an *attitude*? "Attitudes are usually defined as a disposition or tendency to respond positively or negatively towards a certain thing (idea, object, person, situation). They encompass, or are closely related to, our opinions and beliefs and are based on our experiences."[1] Often when I was concerned with the attitudes of leaders who worked for me, I asked them about their past and for whom they had worked. Usually I discovered that their experiences and the examples that the people who they worked for were responsible for their attitudes and leadership styles. Learning this, I was able to work with them to reshape some of their attitudes.

Does your attitude really impact others? The underlying premise of a host of self-improvement books tells us that the answer to this question is yes. Books such as *Success Through a Positive Mental Attitude, The Power of Positive Thinking,* and *Think and Grow Rich* preach that the results of our attitude are evident in ourselves and those around us.

Your attitude determines where you will go much more so than your aptitude. Employees are more concerned with how you think and how you treat them than they are about how much you know. Employees will initially assume that you have the aptitude for a job if you were selected for it. However, your attitude toward your employees will determine whether or not you are successful. Have you ever worked for someone who everyone was glad to see leave — and in fact, the party was held after the person left?

How does the attitude of your company, or supply chain, or representatives of your supply chain impact customer attitudes and perceptions of your company? Here is a short case study of one major company and its customer service:

My cell phone was "on the fritz." A problem that no one wants to have happen because over the past ten to fifteen years, cell phones have gone from a luxury to a necessity to stay connected in today's society. I took my phone to the local XXX store and repair facility as advertised on the company's Web site. The Web site listed the hours of service as 8:00 a.m. to 5:00 p.m., Monday through Friday. I arrived at the facility at 4:30 p.m. only to find that the only authorized repairperson was not there. I inquired about the hours and was curtly told by the manager while pointing to a sign, "Repairs are from 8:00 to 5:00." I was then told, "Things were slow so I let him go home early. Come back tomorrow." Because this was not an option, I, after having a problem with the service from this "independent contractor," contacted the parent company's "customer service line." The representative in the call center in Florida (I and my phone were in Kansas) told me to take the phone to the repair center that I had just left and then offered two alternative locations not even close to my house. When I explained that those locations were not close, I was told, "Well, I am in Florida so I don't know what is close to you."

After a run-around on the customer service line, I contacted the parent company via its customer service link on the Web site. Figure 6.1 shows the response I received two weeks later. I guess the company must have thought that its third-generation phone was so good that it fixed itself.

This company insisted that their "authorized representatives" were merely independent contractors and not representatives of the company. This "independent contractor" displays the sign of the parent company in the window, and uses the trademark symbols of the company on its Web site and in its advertising. When it is convenient, this company with links to the Web site of its "authorized representative" refers customers to the store of its representatives; but when there is a problem, these representatives become "independent contractors." How do you deal with your supply chain partners? Are they the face of your company or merely "independent contractors?" How do your customers see your "independent contractors" using your logo and Web site? What behavior are you modeling for your customers to see in respect to your supply chain partners?

Subject: Email response delayed–Please contact us if you still need assistance!
From: X Company Customer Relations–eCare
Date: Friday, April 4, 2008 9:10

*"Company X recently experienced a system issue that has
prevented us from reviewing and responding to your recent
e-mail. Although we have been able to resolve our system issue,
I am concerned that you may still have questions.*

*If the needs of your original message remain and you require
assistance, please respond to this message and a Company X
representative will be happy to help you. Please accept my
apology for any inconvenience this may have caused you.*

*You are a valued customer–We value your business and
appreciate your understanding.*

Sincerely,

XXXX
Director eCare Operations
X Company

Figure 6.1   Customer care e-mail.

What attitude do you show concerning your customers? Your true character is what is displayed when you think no one is watching or can overhear what you are saying. This was recently overheard in an office in May 2008: "What do you mean that they won't do it? They are just customers. Who do they think they are?" As I looked around to see if I was the only one who that heard this comment, I saw a couple of the company's customers only a few feet away shaking their heads. Apparently, they heard the same comment. Do you think they will be back as customers in the future?

Is there a link between your attitude toward your employees and how your employees treat your customers? Sam Walton believed that your attitude toward your employees would show up in the attitude that your employees will show your customers in less than two weeks. Try it for a couple of weeks and see what happens. The results may surprise you. What do you have to lose?

Can the attitude of one employee impact the entire company and the company's product line? The *Philadelphia Inquirer* (May 21, 2008 issue) had an article on the front page by Maria Panaritis and Stacey Burling entitled "Arrest Made in Chinook Vandalism." Apparently a disgruntled employee severed the wires in a "fire-hose-thick bundle of wires" of two Chinook helicopters. All the investigators could provide was that the offender "might have been unhappy at work." The only explanation from the union representing the employee was that several employees were being transferred to another building to work on another project — apparently this employee did not react very well to the change. This act cost thousands of dollars and shut down an assembly line. What could a disgruntled employee do to your operations? Can you shape the attitude of your employees by modeling the right attitude for them to emulate? Absolutely, a little extra effort by the first-line supervisors and other company leaders could have shaped the attitude of the employee and helped him to see the benefits of the change. Or, at a minimum, walking the process on a regular basis and talking to employees would have identified this attitude problem long before it impacted the program and cost the company thousands of dollars.

> Don't judge. Even God waits until the very end.
> —Harvey Penick, famous golf teacher and coach

Do you judge people based on their color, religion, or appearance? This reflects your attitudes. Attitudes toward people based on the color of their skin are what caused and still cause problems in most countries. In the United States, the attitude toward African-Americans and people of Latin descent led to players being banned from playing baseball in the "white" Major Leagues. This led to the formation of the Negro Leagues in 1920.[2] Even in the U.S. Armed Forces in the twentieth century, there were segregated units. The attitude of prejudging people based on color or religion still exists. The prejudging of an entire religion led to the "Final Solution" by the Third Reich in World War II that is now simply known as the Holocaust.[3]

Sometimes it is not just a person's color or religion that causes us to judge them — it is simply their appearance. In the movie *Pretty Woman*, Julia Roberts' character experienced this when she went shopping on Rodeo Drive. In real life, my Aunt June had a brother who owned and ran a very successful automotive garage in Nashville, Tennessee. One day, Uncle Allen went into one of West Nashville's more exclusive dress stores to buy an anniversary present for his wife. He was not dressed like the normal client of the store. This caused a little tension in the store. When Uncle Allen told the clerk what he wanted, she replied, "Do you know how much that costs?" Uncle Allen simply replied, "I don't really care unless it is more than $3,000 because then I would have to get some more money out of the truck." The sales clerk assumed that Uncle Allen did not have the money because he was dressed in overalls. The moral of this story is much like what we have all been taught since grade school: don't judge the book by its cover.

William James once said, "The greatest discovery of any generation is that a human being can alter his life by altering his attitude." Conversely, how much can you change your company or organization by altering your attitude and the attitude of your employees? One of the greatest discoveries of this generation of business leaders is that you can change the life of the company by changing the way you think.

Is there a link between your attitude as a leader and the onset of motivational dysfunction in your employees? Absolutely! Here is an example of how your attitude can impact your employees. If you constantly berate your employees, how can you expect them to develop a healthy attitude about themselves? If you do not create a positive environment for your employees and show a good attitude, how can you expect the morale and attitude of the employees to be good? You have to model the right attitude for your employees. They can catch a good attitude or a bad attitude from you — they are both contagious. Which one do you want to spread? The choice is yours.

Just as your attitude impacts those around you, do you let the attitude of others impact your attitude? And if you do, do you then take out that attitude on your employees? I have often asked audiences this question. I then tell them that experience shows that when you allow the attitude of others to negatively impact your attitude, you then take out that negativity on your employees and the chain continues down the chain. At the end of the chain is a frustrated employee who goes home and slaps the dog or cat. Sometimes, instead of ruining everybody's day, it would be easier to pick out an employee and go slap his/her dog or cat. Just making a point here; please do not think that I approve of slapping a dog or cat — in fact, probably the most soothing thing in the world is the lick of my puppy and her wagging tail when I come home.

However, sometimes even I allow the attitude of others to impact my treatment of employees. How often do you take out your attitude on someone else? I had an employee who it seemed like I just could not communicate with and could not seem to get through to. For years I had been telling my employees not to let their attitudes impact how they treated their employees based on watching some of my previous bosses and employees. It seems that every time I received a butt chewing from my boss, this one employee happened to be the next person I saw. Can you guess what happened? Every time I received a butt chewing, she got a butt chewing from me simply because she was the target of opportunity.

It was not until one day shortly after the September 11, 2001, attacks that I was made aware of this situation. I had been called on my day off to go to the railhead at Yermo, California, to investigate an accident. In the haste to unload the railcars, a Bradley Fighting Vehicle was driven off the side of the train rather than off the ramp at the end. Luckily, no one was hurt and, with the exception of a small fuel spill, there was only minor damage. But the frustration was that I had lost a good part of my only day off that month. So who called while I was still hot? Of course it was this one employee, once again with bad timing. After I got off the phone with her, my daughter, who just happened to be standing there with her when I unloaded on her, called and asked why I was always so mean to this employee. I finally realized that I was doing exactly what I had been telling people not to do. I am proud to say that this employee became my teacher and is now my closest friend. Over time, I have learned much more from her about leadership than she has ever learned from me.

Do you create an attitude of customer service in your organization? The Home Depot used to pride itself with the employee attitude expressed on their aprons, "I work in all departments." This used to mean that if you had a question about the location of an item in the store, the employee would take you to the correct store location. During the past five years, this practice has eroded significantly. In fact, in my local Home Depot, you are lucky if you can find an employee at all.

After the retirement of Bernie Marcus and Arthur Blank, The Home Depot went through a total attitude adjustment. In this respect, I am not referring to the "attitude adjustment" offered by many local bars and restaurants. Rather, I am referring to a change in the attitude toward employees and customers. This was the type of attitude adjustment that negatively impacts stock price and customer perceptions. The new management wanted to save money by hiring more part-time employees. Many of these employees were not as dedicated as their full-time predecessors. This, coupled with the arrogant attitude of the new management, resulted in a loss of market share and a steady decline in the stock price. This is a good example of Sam Walton's theory of management attitude and the attitude shown customers.

Do you create an attitude that promotes creativity in your organization? I believe the master of this was Walt Disney. He created an atmosphere that rewarded creativity, and the results are evident in Disneyland, Walt Disney World, and Disney parks around the world. But he is not alone; companies such as Rubbermaid and 3M encourage and reward employee creativity. The results of this attitude can be seen in the new products introduced by these companies on a regular basis.

Or, do you stifle creativity? Do you subscribe to the "We have always done it this way" philosophy? Or worse, do you promote the philosophies of "If I did not think of it, it must not be good" or "My way is the only way"? One of the biggest causes of motivational dysfunction in any organization is the "We have always done it this way" philosophy. What if the way we have always done it is wrong? Some companies have operated wrong for so long that wrong looks right. Not allowing creativity stifles employee initiatives, thus promoting motivational dysfunction. Many companies suffer from "NIH syndrome." If it was "Not Invented Here," it must not be of any value. This is a variation of the "If I did not think of it, it must not be any good" mentality that serves to deter employee motivation and initiative. Do you have an attitude of a closed mind? If so, how can you expect your employees to have open minds and accept suggestions or performance improvement tips?

When talking to prospective clients, I look for an attitude that welcomes change and really wants to improve. When I hear, "We are unique," I usually feel the attitude of skepticism and unwillingness to listen to new ideas and creativity.

Do you have an optimistic attitude? Are you always looking for the good in a situation, or do you have a defeatist/pessimistic attitude? Are you always looking for the bad? You have to look for the good in any situation. When evaluating your performance or the performance of your employees, more often than not there is something that is good and needs to be sustained. Look for that as a way to give the employee or yourself a pat on the back. Look for employees doing something good rather than for someone doing something wrong. This attitude will go a long way in preventing motivational dysfunction.

Do you develop a winning attitude in your organization? As a leader you are a coach. Have you ever watched a sports event and noticed the attitudes displayed by the different teams. When a team is on a losing streak, it is easy to notice how down the players appear. When a team is winning, it is easy to show a winning attitude. The difference between good coaches and bad coaches is the ability to create a winning attitude. In the early 1980s, the Atlanta Braves were coming off a number of "rotten" seasons. The new manager, Joe Torre, stressed winning and, in fact, the Braves had a great spring training and started the season with a winning streak. During spring training, Torre was asked about the spring games. The writer commented that the spring training games did not really matter. Torre replied that winning games in spring training sets the winning attitude that would carry over into the regular season — and it did. Conversely, teams that get on a long losing streak exhibit a losing attitude — and the longer the streak, the harder it is to break that attitude. The same is true in business. If a company has a couple of bad quarters, all of a sudden the attitude of the employees starts to look like that of a team on a losing streak.

As a leader it is your responsibility to instill and model a winning attitude for your employees to emulate. Coaches John Wooden and Dean Smith were masters of creating this winning attitude. Coach Jimmy Valvano had a penchant for creating a winning attitude and used that attitude to create a National Champion out of the 1983 North Carolina State University Basketball Team. Even after developing terminal cancer, Coach Valvano continued to display a winning attitude and only a few months before his death told a national audience watching an awards ceremony on ESPN, "Don't give up; don't ever give up."

What about an attitude of customer service? Do you model a customer service attitude for your employees to emulate? Think about your favorite restaurant; how often do you see the manager on the floor talking to the customers? Think about your favorite store; how do you feel when you see the store manager walking the floor and talking to the customers? I can't answer for you but for me, when I see a manager on the floor — and it could be a store, a restaurant, or a warehouse — I get a feeling that the manager is concerned about me as a customer. How often do you get out of the office and talk to the customers? For that matter, how often do you get out of the office and talk to the employees? Sometimes it helps customer service to see a manager patting a worker on the back. The customer feels like the manager is concerned about his/her employees and is more than likely also concerned about the customers.

What is your attitude toward diversity in your workplace? At the end of World War II, the biggest complaint in the warehousing industry was that employees did not speak English. In today's society this is still a common complaint. What can you do about this? If the predominant language in your operation is Spanish, try learning some Spanish.

Do you teach tolerance and diversity acceptance in your organization? And if the answer is yes, does it come across as "checking the block" lip service or is it a sincere concern about employees from different cultures? Is an attitude of diversity important to success in today's global economy? How do you develop this attitude? One way may be to have diversity lunches that offer employees the opportunity to sample foods from other cultures. Another successful program is to have training programs that highlight the cultures of the different employee backgrounds.

One of the most amazing examples of a lack of tolerance for people from a different ethnic background comes from the Theater Distribution Center[4] in Kuwait at the start of Operation Iraqi Freedom. I was fortunate enough to have been there in person when this incident took place, so it is not from second- or third-hand stories that I relate this incident. It was over three weeks into the operation of the Theater Distribution Center before we had a full-time supply unit to help run the Distribution Center.[5] The Platoon Leader for this supply unit was an African-American (not really important except for the moral of this story). The first contract workers at the Distribution Center were four Bangladeshi forklift operators. These four workers had been there twelve to fourteen hours a day every day since the inception of the Distribution Center, to include the first day of hostilities when the SCUD missile alarms went off eight times. They were hard-working, dedicated employees.

After about two days of working in the Distribution Center, the platoon leader came to me and said, "We need to talk." After moving to a location that was a bit more private than the middle of activity, he stated, "Sir, I cannot work with those people." I was not sure which "people" he was talking about, so I asked him to elaborate a bit. That was when he pointed at my Bangladeshi forklift drivers and said, "They do not speak English and I do not like them."

After a little history lesson and an explanation of the work ethics of the forklift drivers, I asked the young platoon leader if he would be offended if someone were to come to me and say that they could not work for him because he was African-American. I also explained to him that prior to his arrival we had no problem communicating with these forklift drivers, and if he would get off of his "Superior American" pedestal, he would see that he was treating these workers the same way his family had been treated in the past.

The good news here is that after this discussion and a slight look of embarrassment, this platoon leader came around and became a leader of all the workers — not just those who shared his background or homeland.

Diversity is much more than legislated equal opportunity or affirmative action. These programs are dictated by law, whether you like them or not, and your attitude toward these programs really does not matter. What does matter is your attitude toward the growing diversity in the workplace. How do you ensure that people of diverse backgrounds mesh into a workable, productive team?

What about your attitude toward handling adversity? Probably the best example in recent times about handling adversity is the story portrayed in the movie

We Are Marshall! For those not familiar with this story, the Marshall University football team and many of its supporters went down in a plane crash during the 1970 season. The rebuilding of the football team and the healing of the city showed the ability to deal with adversity.

How do you handle adversity? This is linked to your attitude more so than to your aptitude. A good friend of mine, and also my former lifting training partner, had a great outlook on adversity. Lonnie Keyes always told people, "Adversity is my friend." This same attitude enabled him to come back from a potentially career-ending detached triceps tear to bench close to 500 pounds in competition. Lonnie's attitude is similar to the works of Friedrich Nietszche when he said, "Whatever does not kill me makes me stronger." What makes you or your company stronger is not the adversity but rather how you handle it and modeling a behavior and attitude that instills confidence in your employees so you will be able to handle adversity without freaking out or blowing a gasket.

Some people do not handle adversity and pressure well. During the early days of Operation Iraqi Freedom, I was enjoying my first hot meal of the day in the Camp Arifjan Dining Facility when the SCUD missile alarm sounded. This was a good dinner and the timing was not good at the end of a very long day. That morning we had all been informed that, contrary to what had previously been put out that we should don our protective gear and move to the center of the building for safety, we should move closer to the walls in case the high ceiling collapsed. Apparently one of the general officers did not get the word on the change. When the alarm sounded, he started yelling at everyone from the top of a table, calling us a string of obscenities and telling us that we should be in the middle of the room away from the walls. The amazing thing about this whole situation, besides the fact that it showed that this particular general was not handling the stress of the war zone very well, was that the only person in the building who was not in protective gear and protective mask was the general himself. If the missile did hit the building and if it was carrying a chemical warhead, the most severe casualty would have been the person yelling at everyone else. Your confidence in yourself and in your employees determines how you will respond to adversity in your organization.

What about an attitude of arrogance? There is no place in leadership or management for an attitude of arrogance. For some reason, in all companies and all industries, there is a number — thankfully a small number — of individuals who think that they deserve special treatment because of their position or title. This leads to an attitude of arrogance. For some unknown reason, a promotion or new job title seems to make some folks think that they are all of a sudden better than others. One assistant store manager was overheard telling an employee asking for help with customers, "I do not get paid to work; I get paid to delegate." When I was in Kuwait, it was interesting to watch individuals who thought that they should be treated special and were too good to do manual labor or wait in the dining facility line. Because of this attitude by some of my cohorts, soldiers were surprised to see me on a forklift moving supplies or climbing on trucks to see what was on them.

As the Director of the U.S. Army's School for Command Preparation, I had responsibility for training not only future commanders of organizations of up to 8000 employees, but also the opportunity to train their spouses for the challenges of being the spouse of a senior military commander.[6] Often, some of these future commanders would come in with an attitude of arrogance. They had a right to feel special because only a very small percentage of all U.S. Army officers are chosen for senior-level commands. Feeling special is one thing but allowing that to become an attitude of arrogance is detrimental to the morale of the future unit. One of our regular speakers told the groups that for every one of them who was there, there were at least three other extremely qualified officers who were not selected but could easily fill in for them. There is nothing wrong with being proud of your accomplishments as a leader but do not let that pride become arrogance. In professional sports, arrogance and ego often serve like a cancer spreading through the locker room and oftentimes cause a good team to decline in quality. As a leader, you must prevent that in your organization.

There is one other attitude that has no place in leadership modeling. That is the attitude of ego. The Chinese philosopher Chuang Tzu addressed ego when he said, "Ego is the worst thing that can happen to a person." Everyone has an ego. What is detrimental to leadership is to allow ego to get in the way of mission accomplishment and motivating employees.

Can the attitude of one employee impact the entire company and the company's product line? The *Philadelphia Inquirer* (May 21, 2008 issue) had an article on the front page by Maria Panaritis and Stacey Burling entitled "Arrest Made in Chinook Vandalism." Apparently a disgruntled employee severed the wires in a "fire-hose-thick bundle of wires" for two Chinook helicopters. All the investigators could provide was that the offender "might have been unhappy at work." The only explanation from the union representing the employee was that several employees were being transferred to another building to work on another project — apparently this employee did not react very well to the change. This act cost thousands of dollars and shut down an assembly line. What could a disgruntled employee do to your operations? Can you shape the attitude of your employees by modeling the right attitude for them to emulate? Absolutely, a little extra effort by the first line supervisors and other company leaders could have shaped the attitude of the employee and helped him to see the benefits of the change. Or, as a minimum, walking the process on a regular basis and talking to employees would have identified this attitude problem long before it impacted the program and cost the company thousands of dollars.

There is one more attitude aspect that is important in supply chains. You have to have an attitude that says you love what you are doing. I heard a conversation recently between noted experts in the supply chain profession. One commented how refreshing it was to have a conversation with someone who loved the supply chain business as much as he did. You have to love what you are doing; you have to love your job, love your profession, and love the people who you are working with and those who are working for you. You will never be a supply chain leader if you do not love your profession. Life is way too short to be in a business that you do not enjoy, and you cannot model the right attitude for your employees if you do not love what you are doing.

The key to modeling attitude is to adopt an attitude of being yourself. There is only one George Patton; only one Martin Luther King, Jr.; only one Jack Welch; and only one John F. Kennedy. You cannot be one of these leaders. However, you can model your leadership style from how these leaders act and react to stressful situations. But you still must be yourself.

Aptitude

What is *aptitude*? Why is this important for a leader? The *Merriam-Webster Dictionary* defines aptitude as "an inclination; a natural ability; a capacity for learning."[7] It is a little more than that. Your aptitude in today's business is not only your ability to learn, but is also a measure of what you know about your job. Why is this important to leaders? You cannot bluff your way through today's supply chains. Some folks definitely try to, but you have to know what is going on in your operations and also understand what is going on in your operations.

What is more important in today's business world, your aptitude or your attitude? Both are important. Your boss is concerned about your aptitude for the job, and it is most likely your aptitude that gets you hired for a job; however, it is your attitude that will determine what you are from the perspective of your employees and customers.

Your aptitude is reflected in your knowledge, skills, and abilities that are rated and judged in the hiring process based on your professional experiences and education. It is also reflected in your ability to learn new ideas and techniques. Today's supply chain is not what most of us came into when we started in this business. Your aptitude for learning new techniques and mastering new processes will determine your "promotability" in any organization.

Self-development is another aspect of the House of Leadership that we will look at in a subsequent chapter. Self-development is closely related to your aptitude. Your aptitude will get you noticed and will set the conditions for success in your career. Aptitude will also enable you to see where your employees and your organization need to improve in order to accomplish the missions necessary to remain competitive.

Learning about leadership skills is part of the aptitude necessary for personal and professional growth. Studying leadership and studying supply chain management techniques and case studies will help you improve your personal abilities. Studying leadership will not necessarily make you a better leader but it will give you an idea of how other leaders responded, thought, and acted in certain situations. It is the aptitude to apply these lessons that will make you a better leader. You can read all the observations and accounts of leaders in any walk of life or any profession but if you do not try to apply the lessons to your own actions, the lessons are not lessons but simply observations. Your aptitude for applying new concepts will enable you to apply the lessons of leadership and enable you to better lead the people who make up your supply chains.

Are aptitude and attitude related and complementary? Absolutely! Your aptitude in supply chain matters will enable you to get in the door as an order qualifier. However, your attitude will be the order winner that places your company or supply chain ahead of others. You may very well have the aptitude to manage the supply chain but your attitude with your customers will determine how long you get to serve those customers in your supply chain.

Accountability

What is *accountability*? Growing up in the military logistics world meant that you knew exactly where everything was at all times. In today's supply chain world, it means the same thing. You know exactly where every order is in the supply chain and where every item of product is in your facility. The loss of accountability of items of inventory is a direct loss to the bottom line of your company.

However, in the modeling of leadership, accountability means assuming responsibility for your actions and the actions of your employees. It also means assuming responsibility for the actions of your bosses. It is closely tied to ethics.

Why is this important in modeling leadership? Your employees need to know that they can take actions to meet the needs of the customer with the confidence that you will support their actions as long as those actions do not endanger another employee or violate any laws or ethics.

Conversely, leaders must be willing to accept responsibility for the actions of their bosses. How many times have your heard one of your bosses or co-workers say something to the effect of, "The boss said to do this; I do not think it will work but we must follow his directions."? Does this set the conditions for success in the operations? Probably not; and when the operation fails, this same manager usually says, "I told you it would not work." What does this do to build the confidence of the workers for the next higher-level boss? What it does is make the bosses look bad. However, when the boss says to do something and then accepts the blame, even when the workers know it was the guidance of the next higher-up, the employees gain confidence in their boss knowing that he or she will most likely take the blame for their mistakes as well. The key to accountability is that leaders take the blame for the mistakes or actions of their employees but always pass on the praise when employees take a risk and succeed.

Accountability is also a corporate responsibility. Leaders must ensure that their companies are accountable. Companies must be accountable for their actions; they must be accountable for their products and how their products are made; and they must be accountable for their distribution and manufacturing processes. Corporate accountability leads to trustworthiness with employees and customers. There is a direct tie-in between personal and corporate accountability and honesty. When a company is perceived by customers as honest and accountable, they become loyal employees.

Are you accountable for your actions? Do you take responsibility for your actions, or do you try to find someone else to blame? When you take responsibility for your actions, you will start to attract loyal and accountable employees.

Are you accountable to your customers? Do you and your company take responsibility for the products that you manufacture or sell? In 2007, Mattel took responsibility for the products that it had manufactured in China. The big news was that millions of toys and products were recalled. The real news was that Mattel actually showed corporate responsibility by admitting that there was a problem before it became a CNN news story. Unfortunately, the news companies missed the real story.

Are you accountable to your employees and employer? Are you a loyal employee? If you are a loyal employee, then your personal goals will mesh with the corporate goals. If you are a loyal employee, then you will be accountable for your actions, and your loyalty and accountability will promote loyalty and accountability in your employees.

Leadership accountability goes beyond just being personally accountable for your actions and the actions of your employees; it includes fiscal accountability. Are you fiscally accountable to your shareholders? Leaders must be accountable to their employees, their employers, their customers, and their shareholders. When leaders maintain a high level of accountability to these entities, they become trusted and their companies continue to move to new levels of excellence.

Modeling leadership for an organization must include establishing a positive, winning attitude, learning as much as you can about your profession, and remaining current in the field while accepting accountability for your actions and the actions of your bosses and your employees.

Notes

1. http://tip/psychology.org/attitude.html. Accessed October 10, 2007.
2. The Negro Leagues Baseball Museum is located in Kansas City, Missouri only a few blocks from The Paseo YMCA where the official Negro Leagues were formed in 1920. For more information on the Negro Leagues, visit the museum in Kansas City or go to http://www.nlbm.com.
3. If you are ever in the Los Angeles area, visit the Museum of Tolerance to see exhibits of not just the Holocaust and the persecution of Jews, but also incidents of discrimination and attitudes based on race and religion. The Museum of Tolerance is collocated with the Simon Wiesenthal Center in Los Angeles.
4. The Theater Distribution Center was the first distribution center established in an active theater of war since World War II. There were military and commercial operations at Cam Rahn Bay in Viet Nam but this distribution center in Kuwait was the first inland distribution center in war in almost 60 years.

5. As written in *The Forklifts Have Nothing To Do!*, the first several weeks of the Theater Distribution Center the soldiers who worked in the Center were borrowed military manpower (temporary workers). We had a new crew of "volunteers" every 12 hours to get the Center established while we waited on the arrival of a trained supply unit.

6. The demands on military spouses are much greater than those placed on the spouse of a leader of a similar-sized organization in the civilian community. The spouse of a military commander is not a paid position but includes hours of volunteer work leading family support groups, comforting other spouses, and serving as the focal point for information when a unit is deployed. And unfortunately, the spouse of a military commander may end up helping break the bad news when a soldier is killed or injured in combat. Because of these special needs and duties, the U.S. Army pays for the spouse to come to Fort Leavenworth for a week of specially developed classes on team building, counseling others, and how to set up a successful family support group prior to unit deployment to make sure the spouses are taken care of when the unit is away for a long deployment.

7. http://www.merriam-webster.com/cgi-bin/dictionary. Accessed October 10, 2007.

Chapter 6 Questions

1. What is your attitude saying about you that you may not be aware of?
2. How do you deal with an arrogant supply chain provider?
3. Does your attitude toward your customers really reflect your respect for the customers?
4. Are you accountable to your employees, your company?

Chapter 7

D5
Determination, Dedication, Discipline, Devotion, Decisiveness

D5 — this is not the next sequel to Disney's *The Mighty Ducks*, although some of these attributes are evident in the team that comprised The Mighty Ducks. D5 is the next pillar in building the House of Leadership. D5 is determination, dedication, discipline, devotion, and decisiveness. D5 includes the attributes of the determination to set personal, professional, and educational goals; the dedication and discipline to achieve these goals; the devotion to your employees to assist them in achieving their goals; and the decisiveness to make the hard decisions.

Why is goal setting important? From a personal perspective, it is important to set goals for both your professional life and your personal life. Setting goals provides you with a road map of where you want to go and how to get there. The old cliché "If you don't know where you are going, you will not know it when you get there" is true for setting goals. If you do not have goals, how will you determine if you are successful? What your definition of success is will determine some of your goals.

Where do you want to be in the future? The book *Think and Grow Rich!* sold millions of copies using the concept of developing your definition of success and setting your goals to reach success. In supply chain leadership it is just as critical to set goals for your supply chain as it is to set personal goals.

Here is an example of setting and achieving personal goals: I wrestled at 105, 112, and 119 pounds in high school and at 119 pounds my freshman year in college. My goal in wrestling was simple: I did not want to get pinned. My goal was never to win the state championship or even the district championship; I simply did not want to get pinned. When I finally made it to the regional competition in my senior

Figure 7.1 World record exceeding squat; 840 pounds at a bodyweight of 196 pounds.

year, my goal was still the same: I did not want to get pinned. I went to the regionals knowing that if I did not win the first day, I would have to pay for my hotel room. I ended up paying for my hotel room but did not get pinned. I set low goals and achieved them. What goals are you setting for you and your supply chain?

It takes determination and proper goal setting to go from wrestling at 119 pounds to setting a world record in weightlifting at a bodyweight of 196 pounds (Figure 7.1).

When I stopped wrestling, I decided I was tired of being the smallest person in the school and dedicated myself to gaining weight through serious weight training. This time I set a series of intermediate goals and some specific long-range goals. My first foray into competition was a result of these goals. My intermediate goal was to get big enough with enough muscle definition to compete in a bodybuilding competition. In 1977, I competed in the Mid-South Bodybuilding Meet in Durham, North Carolina. After competing in my first competition, I set a new goal. My new goal was to *place* in a bodybuilding competition. By March 1978, I achieved that goal. My long-range goal was to *win* a bodybuilding competition. I continued to compete and started placing in more competitions and then in 1979 won the Army Hawaii Bodybuilding Championships.

While training for wrestling and then for bodybuilding, I realized that I also had a talent for powerlifting. Bodybuilding and powerlifting are not always complementary sports because bodybuilding requires dieting (something I really got tired of while wrestling) and a period of high-repetition lifting to define the muscles, while powerlifting requires no dieting and fewer repetitions to develop strength. While still in college, after reading an article on the Hawaii International

Powerlifting Championships (HIPC), I told my training partner that I would compete in that meet. I had already placed well in some local competitions and set my goals on a higher level. My training partner just smiled. I took that as a challenge and it strengthened my resolve to compete in the HIPC. In 1982, I not only competed in the meet but placed second. My goal was now to win the meet. In 1984, I did just that. In the meantime, I set a few more goals. The first was to win a national championship; the next was to set a state record; and the biggest goal was to break the world record for my weight class in the squat (see Figure 7.1).[1]

These goals required dedication, determination, and self-discipline to reach. During the course of my twenty-year lifting career, I accomplished all three goals. I set over seventy records at the state, All Army, and Armed Forces levels in five different states and on two continents. In 1987, I squatted 840 pounds at a bodyweight of 196 pounds and broke the existing world record by fourteen pounds. In 1992, I achieved the goal of a second national championship at a bodyweight of 198 pounds. By the time I "retired" from competition, I had set more than sixty Armed Forces records, state records in seven states, exceeded the world record in the squat, and finished the 1987 competition year ranked first in the world in the squat and tenth in the world in total weight lifted. In addition to the HIPC, I won two national championships, multiple state championships, an Armed Forces championship, and two European Armed Forces championships. All of this was a result of dedication, determination, devotion, and discipline.

The traditional acronym used in goal setting is SMART. When I first heard this, my thoughts were of course that I am going to set *smart* goals. After all, who would set *stupid* goals? Sometimes I really wonder if some folks do set stupid goals. But the SMART acronym does not refer to smart or stupid goals; it refers to setting goals that are:

Specific in nature. When assisting employees in developing performance goals for the upcoming year, many leaders and managers are adept in setting very specific performance goals. However, when it comes to setting personal goals, these same managers and leaders are not always as adept in setting specific goals. Just as employee performance goals should be specific in order to produce the desired outcome, personal goals need the same specificity. A generic goal is much like the cliché of not knowing where you are going and when you get there.

When it comes to setting supply chain goals, leaders and managers are often guilty of setting goals that are not specific in nature and therefore are perceived by the employees as unachievable or so nebulous that the employees are not sure what to do to meet the goals or objectives.

Measurable. We all know that we can only manage those things that we can measure. The same is true for goals. It is important for supply chain leaders to set measurable goals for their supply chains and for the personnel who make up the supply chain. How many people want to be evaluated on their annual appraisals on goals that are not measurable? How will you know if you met your goals if there is no measurement for them?

Setting goals for your supply chain that cannot be measured will result in frustration for all involved. And if they are not measurable, how will you be able to tell the boss that you achieved them, or worse, how will you be able to convince the financial people to fund the projects that support the goals? It is a fact that financial departments are setting measurable goals for your projects; if they can, why is it so difficult for supply chain leaders to quantify their projects or goals?

There is a warning here. Be careful about being too optimistic on your projections for your projects. Several years ago, a young staff officer working for me on the U.S. Army's Supply Chain Process Improvement Program made a boastful claim that this one project would save the U.S. Army several million dollars in the next year. The financial staff heard these claims and immediately removed the money from the logistics budget. Luckily the project was more successful than the project officer imagined and the impact was minimal. Be sure when you quantify your projects that you have done your homework. If you are going to "save" money for the company based on the requested budget, make sure your measurements and analysis are correct.

Achievable and Realistic. When I was setting goals for my lifting career, I set goals that were achievable and included intermediate goals for each major goal. When setting personal goals, it is a little easier to set goals that are perceived as achievable. In fact, the tendency is to set goals too low. It is important to have intermediate goals and stretch goals that you have to work hard to achieve. These stretch goals require dedication and determination to set realistic goals and then require the devotion and discipline to reach them.

The same is even truer for supply chain goals. Everyone in your supply chain must perceive your supply chain goals as achievable if you are going to be successful in achieving them. A supply chain leader may set goals that he/she believes are achievable but if his/her employees do not perceive them as achievable, the company will be a failure in achieving the supply chain goals.

Here is an example of setting goals that were not perceived as achievable by the people who had to achieve them. In 1995, the U.S. Army started a Supply Chain Process Improvement Program known as the Velocity Management Program. Part of the strategy of the program was to establish performance improvement goals for customer wait time reductions.[2] To add teeth to the goals set for the improvements, the Vice Chief of Staff for the U.S. Army[3] sent out the set of goals and objectives at the beginning of each new fiscal year. In the first several years the goal for customer wait time reductions was the same for the Active Army (the Active Army consists of those soldiers that have volunteered for the Army and are full-time soldiers) as it was for the U.S. Army Reserves and the Army National Guard (the U.S. Army Reserves and the Army National Guard soldiers at that time trained one weekend a month and for fourteen days of Annual Training once a year). There were some major differences between the Active Army and the Reserve components in 1995 to 1997; these differences have pretty much disappeared since September 11, 2001. But at the time that the goals were being set, the Active Army had soldiers on duty 365 days a year and the Reserve component soldiers were on duty approximately 36 days a year.

However, the goals established by the Vice Chief of Staff's annual memorandum established the same goals for the Active Army as for the Reserve components. The Active Army for the most part perceived the goals as achievable and started working toward meeting the goals. The theory was that as an Army of One, we should have one standard. Most of the Active Army units achieved their goals for customer wait time reductions through a series of process improvements and the establishment of customer-focused metrics. The Reserve components perceived the goals as unachievable and, in fact, their customer wait times went in the wrong direction and actually became longer in many locations.

In 1997, we were successful in convincing the U.S. Army leadership that because the two components of the Army were so different in the available days to process receipts and customer orders, that there should be a different set of goals for each distinctly different component. The result was that the new separate goals were perceived as achievable by the Reserve components and within six months, these units had achieved the goal for the year and came back asking for a new goal. Why? Simply because they perceived the new goals as achievable and worked hard to achieve them, whereas before they perceived the goals as unachievable and therefore did not even try.

Regardless of what goals or metrics you are trying to achieve in your supply chain, the workers must believe that those goals are achievable before they will buy into them and work to meet the new standards. If your current total customer wait time is twenty-plus days, setting a goal for the month to meet the industry average of eight days will probably not be perceived as realistic and achievable. To perceive goals as achievable, they must be perceived as realistic and there must be a strategy to get to your new destination.

Time based. To achieve goals, there must be a timeline assigned to the accomplishment of the goal. A goal without a timeline for accomplishment is like a boat without a rudder. Attaching a time to the completion of the goal is what makes it achievable and realistic. Setting goals for your supply chain must be time based.

Another aspect of time basing goals is to set short-term goals, intermediate goals, and stretch goals. When setting the time frame for goal accomplishment, be realistic. My goal to become a national champion was not an overnight goal. I was cognizant of the fact that such a goal required discipline, dedication, and time. My wife will tell you that my goal accomplishment became somewhat of an obsession — as much as I love the beach, I would avoid the beach until after the national competitions every year. But she will also tell you that my focus was total and my dedication was complete. The year that I exceeded the world record in the squat, I was training in a gym that for some reason did not have air conditioning — almost a necessity in Central Florida in the summertime. The heat did not affect my focus. While training in the squat, I smashed my little finger between the bar and the squat rack, resulting in the fingernail being ripped off. My dedication to doing every planned exercise and repetition allowed me to focus on the workout and the repetitions, and only after the workout was over did I take the time to worry about the finger — probably only then did I worry about it because I was concerned about the impact on the next day's bench workout.

Setting Supply Chain Goals

The D5 is particularly applicable to supply chain goals. As a supply chain leader, you are responsible for establishing goals for the success of the supply chain. This includes modeling the discipline necessary to accomplish these goals. The goals of the supply chain are established through the use of supply chain metrics.

What metrics should you establish, measure, and achieve for your supply chain? And from whose perspective should the metrics be measured? Part of the metrics development process is *benchmarking* your operations. The first step in benchmarking your operations is the admission that there may, in fact, be someone else who actually does the operations better or makes a better product or service. As you benchmark your supply chain, it is important to benchmark your supply chain against similar operations. Unless your supply chain is in the computer assembly and distribution business, you probably should not be benchmarking your operations against Dell. You may want to look at how Dell does business for reference but to set your metrics to compare progress against Dell will be a major mistake and prove very frustrating. As this chapter was being prepared, Dell made the announcement that it was completely revamping its supply chain and getting out of the make-to-order computer business that everyone held as the model of supply chain excellence. So, perhaps everyone now needs to reevaluate their benchmark for supply chain excellence.

The metrics for your supply chain depend on your products and where the products are manufactured and stored. Although Southwest Airlines was able to go to NASCAR to observe pit crew operations as a benchmark for turning planes around on the ground, as a supply chain leader you should be looking at operations closer to your business models.

Questions constantly come into the Supply Chain Research Institute on the measuring and benchmarking of supply chains. What metrics should I use? So and so company uses x number of turns; is that right for my company?

Metrics

Your supply chain vision sets the goals for your supply chain. As a supply chain leader, your vision when properly communicated will provide the motivation to the individuals who make up your supply chain operations to reach new levels of excellence.

Where do you want your supply chain to be? How do you plan to get there? How do you define supply chain success?

Discipline and Determination

Discipline is doing what is right as you work to achieve your goals. The newspapers and the Internet are full of stories about "leaders" who took shortcuts and about athletes who took shortcuts to get to the top. Discipline means working to accomplish your goals while not taking any shortcuts and doing what is morally and ethically right to accomplish the goals.

Determination is the attribute that allows leaders to continue to strive for supply chain excellence despite setbacks or pitfalls. Once you establish supply chain goals and metrics to track your progress in improving operations and satisfying customers, it is determination, dedication, and discipline that will get you and your operations to the top. No one wants an asterisk beside their name because they achieved the "success" but took shortcuts to get there.

As a leader, you have to set your goals for your professional life, your personal life, your family goals, and financial goals for you and your organization. The question of the leader is what are you willing to give up to reach your goals? Can you really achieve a balance between personal goals, professional goals, and quality family time? Absolutely! However, you have to work at it. All too often in the supply chain world, leaders and managers become so consumed in climbing the corporate ladder, or working to ensure that they do not move down the ladder, that they lose focus of what is really important in life. Ask any soldier coming back from a deployment what is important and almost every one will tell you it is time with their families. It does not take a deployment to a war zone to make one realize that families and friends are important to personal well-being.

Can you really be successful in the office and at home? Can you be successful at both without shorting one or the other? The answer is yes but it takes personal and professional discipline and dedication to do this. Are you willing to sacrifice the family for professional achievement? Some folks are. How many times have you heard someone say, "He would sacrifice anything to be on top."? If you choose to sacrifice the family for career progression, be careful because when the glory of the job is gone and the time to retire or move on comes, will it all be worth it if you have no one to share it with? How many times have you heard someone say, "I wish I had spent more time with my children when they were growing up."? I once heard a prominent supply chain leader tell a class of rising leaders, "Be careful what you set as your goals. I had all of the right jobs, all of the critical assignments, and reached the level that I aspired to be and now as I prepare for retirement, I am old and alone. I destroyed two families in my quest for success." Is the price worth it? Assess your personal situation and decide where to place your dedication and determination.

As a supply chain leader, you are responsible for helping your employees set their goals and giving them the tools to be successful. Part of this toolbox is modeling for them the behaviors that will enable them to be successful.

If you preach that they must have a balance between their family and the job, you must model that behavior for them. All too often, employees see the way their bosses are putting in longer and longer hours, sacrificing families and health, and even though they hear what the boss is saying, they see what the boss is doing. This leads to employee turnover and employee frustration. I have heard young, upcoming leaders in several businesses and in the military comment that if what they see of their leaders is what they can expect in the future, then maybe a change of careers is a better choice than what they are modeled by their leaders. Employee retention is a benchmark of leadership — not words but modeling the behavior for them to see. Leadership is a quality that must be seen to be believed.

Another aspect of discipline is knowing what is important and what is just urgent. Everything cannot be a priority. If everything is a priority, then nothing is a priority. In supply chains, everything cannot be the top priority. One interesting revelation that came out of the U.S. Army's Supply Chain Process Improvement Program was a realization that everyone was requesting items at the highest priority level. The result was that items going to the wholesale level at the Defense Logistics Agency depots were all treated the same. Some customers get higher priority treatment than others based on their volume and their contribution to the bottom line profits or reputation of the supplier.

The same is true for supply chain reports. Some reports are always a priority but when investigated are really not even read by the front office. These reports are always urgent priorities but may be unimportant in the overall scheme of operations. When I was at the National Training Center, one of my subordinate commanders told me that there was a series of reports that my boss had to have on a weekly basis. These reports numbered approximately a hundred. Knowing that I did not have time to read all those reports and having a good feel for what my boss thought was important, I held all but four of those reports at my level to see if they were being read or if they were really important. In the following two years, only one of those other ninety-six reports was requested by the boss. The reports were taking lots of man-hours to produce, were marked "important," but in the end were not even urgent. How many reports do you have that are considered "important" or "critical" but in fact are not even urgent — or worse, not even read? Part of the modeling of supply chain leadership is to determine what reports and information are critical and important, and what reports do not add value to the company or the operations.

Discipline and determination work together to establish written goals. There is a power in the written word; and when employees see the vision and goals of the supply chain in written terms that they understand, they will have the personal discipline to work toward those goals. There is something about putting something in writing that gives it credibility and authority. Think about the checkout times of hotels. By placing the checkout times and check-in times in writing, hotels have authority. If you have ever been to Las Vegas and noticed the lines at the checkout counters at 11:00 a.m. or 12:00 noon, you will see the power of putting something in writing.

The same is true for goals. Putting the goals in writing, whether personal, professional, or corporate, gives the goal authority and power. Putting the goals in writing in terms that employees can quickly decipher "What's in it for me?" gives the goals more authority and will empower the employees to achieve those goals.

Where do you want to be? How are you going to get there? Why is this important? If the goals are not perceived as attainable, no amount of discipline and dedication will be sufficient.

Stretch Goals and Long-Term Goals

Just as short-term and long-term forecasting is important in the supply chain, so are short term and long-term supply chain goals. In the U.S. Army's Supply Chain Process Improvement Program, we established short-term goals for each of the aspects of supply chain management on which we were focusing. These short-term goals focused on improvements by fiscal quarters. At the same time we established long-term goals for each fiscal year based on the baseline performance and the quarterly goals. The key to establishing long-term goals against a baseline performance is to keep the original baseline as the standard against which to measure progress. I have seen companies that only measure performance against the previous quarter or previous year and not against the process improvement program baseline.

How do you determine what areas need improvement and where goals should be set to improve the performance of your supply chain? The quickest and most effective technique to identify areas for improvement in any operation is an extension of the guidance of Sun Tzu — get on the ground and see what is going on. Once you are on the ground, ask questions of the employees and develop a process map of all the operations and activities.

Prior to establishing goals for your organization, you have to know the processes of the organization. All too often, leaders and managers set goals for their organizations without really understanding the processes, and the goals or vision for the organization causes confusion in the workforce.

There are numerous methodologies for producing improvements in an organization. The key is to make sure that you are producing improvements and not simply change. In one organization that I worked with, a new "leader" came in and was so jealous of the success of his predecessor that he decided that whatever had the old boss' name on it must be changed. These were not improvements to the organization but simply changes. It became so obvious that this was occurring that the employees started telling the new boss, "Mr. X would not do this way," knowing that the new boss would do just the opposite; they did this as a way of getting the boss to do the actions that they thought should really be done in the organization.

Six Sigma, Lean, and more recently the Supply-Chain Operations Reference-model (SCOR) are the most effective methods for bringing about improvements in the supply chain. Each of these require discipline and dedication from senior leaders, require leaders to walk the process to understand the operations, and require leaders to identify processes that are value adding and those that are non-value adding.

What is a *Six Sigma supply chain?* A Six Sigma supply chain is measured from the perspective of the customer and continues to remove variability from the supply chain to improve customer support. This requires discipline and determination because it often means creating change in an organization. Most employees and a majority of managers are resistant to change because of the fear of the unknown and the perception that change means losing jobs. Experience shows that process improvements should not result in lost jobs but rather a more efficient use of the employees.

Six Sigma uses the Define, Measure, Analyze, Improve, Control (DMAIC) methodology to remove the supply chain variability. The goal of the Six Sigma supply chain is the concept of Perfect Order Fulfillment. The first step in the Six Sigma application to supply chains is perhaps the most important. Defining the process or supply chain is to answer a few questions:

1. *What does the customer want?* Failure to properly answer this question is the major reason for supply chain failure. Anecdotally, every customer wants everything and they want it now. This is the theoretical approach. The military theorist Karl von Clausewitz wrote in his seminal work, *The Art of War*, that all things change when you move from theory to reality. The reality of what the customer wants is important to supply chain design.
2. *What can we do for the customer?* The answer to this question is just as critical as the first question and hopefully matches what the customer wants.
3. *How can we meet the customers' needs better than the competition?* The answer to this question should be your core competencies — those things that you can do better than the competition and those things that serve as your order winners when competing against the competition for new supply chain business.

To answer these questions requires a series of processes: (1) process mapping, (2) benchmarking, (3) metrics, (4) leadership skills, and (5) the application of discipline, dedication, and determination.

A *process map* is a workflow diagram to ensure total understanding of the processes. The process map cannot be developed without walking the process. This is another application of the guidance of General George Patton when he said, "No effective decision was ever made from the seat of a swivel chair." No effective process map was ever made without leaving the office, getting on the ground, and walking the process. The University of Minnesota Web site states that "process mapping is the single most important visual display that can represent 'who does what' within a workplace."

In developing an effective process map, the first thing is to know where/when a process starts and ends. Once a defined start and end point is defined, it is imperative that the mapper walks the process to determine the actual steps/processes. This is not unlike the *Harvard Business Review* (July–August 1992) classic article entitled "Staple Yourself to an Order."

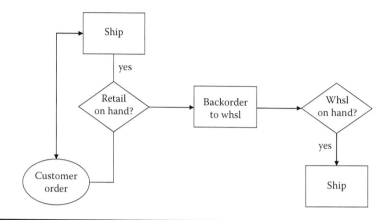

Figure 7.2 Process map example.

As the process is walked, the process map is developed by documenting the sub-processes or steps within the process. To make the map more effective, once the map is developed, add time frames and time measures to the map.

The process map then becomes a visual representation of the work being done in the supply chain, who the customers of the processes are, when the actions take place, and how long each action or sub-process takes (Figure 7.2).

Once the process map is complete, it becomes a training aid for new employees to learn the processes of the operations. Here is where dedication and discipline are important. Once the process map is complete and the bottlenecks, constraints, and value-adding/non-value-adding actions are identified, discipline and dedication are necessary to keep the process map current.

I recently spent time with a major organization on the West Coast, assisting them with their strategic planning, identification of KPIs (Key Performance Indicators), and metrics development. I could not get in to see the CEO until late in the final week of the contract and decided to check out their distribution operations one more time because I had a free day. I noticed that they had a large, very colorful and detailed process map of their receiving operations posted at the inbound dock area. After watching their receiving operations, I realized that the process map was not an accurate depiction of what I was observing. I drew up a process map of the operations and showed it to the CEO the next day. I was quickly told, "That's not what we are doing. Where did you get that?" He then showed me a copy of the process map that was on the wall. When I explained that I spent the day with his receiving personnel observing their actions and that what he had was not the current actions, although certainly more efficient than the reality of what was going on, he agreed to go back and look at what was really happening and was told, "Maybe that is why our numbers are going the wrong way."

Discipline and dedication will keep the process map current and will enable new employees to quickly look at the map and learn the processes and enable them to ask questions about discrepancies between the process maps and the processes they are taught.

The first benefit is that you will be able to identify those activities that are non-value-adding processes. Every process and action in the plant or the supply chain adds cost to the overall process, but not every action or sub-process adds value to the operation or product. The process map will assist in identifying the value-adding and non-value-adding activities. The guideline for value adding is simple: does the activity add value to the product; does the activity add value to the quality of the product; does the activity add value to the bottom line of the company; and does the activity add value to the customer experience? If any of these questions result in a no, there is no value added to the overall process or product. The question then becomes: why do it?

The second benefit of the process map is that it helps you identify the areas that need improvement. Once an area is identified as needing improvement, it is important to quantify and rank order the improvements based on cost and value to the customer or customer experience and then establish goals for improvement.

Time Dedication and Time Discipline

Dedication to the job includes time management. Everyone is given the same amount of time each day. Every one of us gets 86,400 seconds in every day. Yet some people seem to never have enough time for what is important. Dedication as a supply chain leader includes the dedication to efficiently managing your time and the time of your employees.

How many times do you feel like you have accomplished absolutely nothing after a grueling day at work? Have you ever wondered why some people seem to have free time and you don't? Do some people seem to get more accomplished with less effort?

I was recently talking with a colleague about the number of requirements that she needed to get done every day and her frustration over not getting to do everything on her schedule every day. Her biggest complaint was not enough time in the day to do the things that she wanted to accomplish. I asked to look at her calendar. The job that she had was one that I had a few years earlier — so I had a feel for the demands of the job and what was "important" when I had the job.

What I discovered was that she had become a slave to her calendar and further questioning also revealed that she was also a slave to her e-mail (she receives over 200 e-mails a day, usually on her Blackberry®). These are not uncommon problems for leaders and managers at all levels. How much time do you spend every day tied to your computer or Blackberry®, reading and responding to e-mails? And how many of these could be handled by one of your subordinates? Are you delegating responsibility to your employees?

Some of the e-mails and appointments that my colleague was receiving reminded me of Oncken's article in the *Harvard Business Review*, Who's Got the Monkey," or the book entitled *The One Minute Manager Meets the Monkey.* The premise of the article and the book is that if you let them, others will put their monkey on your back and weigh you down. A large number of the meetings on my colleague's calendar could have and should have been pushed down to her subordinates to run and/or attend. A larger number of the e-mails fell into this same category. She was allowing her subordinates and co-workers to put their monkeys on her back.

> I'll pass on this little bit of advice. No matter who you are, you've got to take the time to hug your wife, hug your kids, you've got to take time to pray, you've got to take time to invest in friendships, you've got to invest in other things besides your career.
> — Darrell Waltrip[4]

Who is in control of your professional life? Is it your calendar, full of meetings that end with no other decision except to have another meeting? Is it a set of meetings that the attendees are not prepared for? Do the people at the meetings talk at length because they like the sound of their own voice? Do you spend a large portion of your day reading and responding to frivolous e-mails? If you answered yes or maybe to any of these questions, keep reading.

The one area that we all are given equal amounts of a resource is time. Each of us has twenty-four hours deposited in our bank accounts every day. This account is truly a revolving account. So why do some folks appear to have extra time on their hands and others never seem to be able to find enough time?

Early in the twentieth century, Charles Schwab reportedly earned a large fee ($25,000) from Andrew Carnegie for revealing the "secret" of the "TO DO" list. Your calendar or appointment book is not necessarily your "TO DO" list. Microsoft Outlook is not your "TO DO" list. The first habit that you must adopt to maximize your work hours, and therefore maximize your time to relax and enjoy time with your family or on your hobbies, is the "TO DO" list. I have found that preparing the "TO DO" list and prioritizing the list every night before going home helped me focus my energies for the next day. Waiting until the next morning always proved to be a mistake because someone would come in and interrupt my prioritization time. Once the "TO DO" list is complete, prioritize the items that must be accomplished. When you come in the next morning, start working on the number-one priority and work your way down the list. Prioritizing the items on the list focuses your efforts on what is important and not what is the easiest to accomplish. This activity also allows you to identify items on your calendar that are time users but not productive.

An example of a time consumer that is not necessarily productive can be found in almost every company. This non-value-added time consumer is the routine meeting every morning just to focus your people for the day and then having an evening meeting to review the day's activities. Are both meetings necessary? Would a little time discipline and dedication give you back more of your time everyday? There is only one way to find out. After you prioritize your list, take a close look at the meetings on your calendar — are they productive? Are they necessary? Are they important? Do they have to occur at your level, or can you get by with a summary of the meeting from a subordinate?

The next habit is e-mail discipline. It is easy to become a slave to e-mail. This is even easier with the advent and growth in the use of the Blackberry®. Several years ago, 3M drew a line on its organization chart and monitored the time that executives above that line spent on e-mail. After the study, they limited the time that these executives could use e-mail. They also encouraged the use of telephone conversations and written notes/letters to help the executives maintain their formal communications skills. The habit that I found worked well for me in high-pressure jobs was to set a time to review e-mail. I used to arrive early in the morning before the majority of my employees to read and respond to e-mails; I would check the e-mail just before lunch; and then not again until it was close to time to go home. If a large file came in, I would hit the print button and have it waiting for me when I came back the next morning. This forced my employees to call me if something came in that was hot, rather than assume I read their e-mails. You will be surprised how much time this will free up in the course of the day. Anyway, how many times has someone called to see if you received his/her e-mail. If it is important enough, they will call.

A prominent FORTUNE 500 company went as far as drawing a line across its organization chart and making a policy that those employees above the line would use the telephone and written notes to communicate with people if they could not communicate face to face. Their rationale was that the senior employees were spending vast amounts of time on e-mail and not accomplishing their stated missions.

The third habit to improve your personal productivity is to get out of your office on a regular basis and talk to your employees. What you will find is that problem-solving time decreases because you are getting better, unfiltered information; and you will have a better feel for the work being done and who is doing it. Do not depend on printed reports and e-mails to determine what is or is not being done or produced.

There is another aspect of discipline that should be addressed and that is discipline as a way of helping employees improve their performance. All too often, discipline is used as a method of punishing an employee. There have been volumes written on disciplining the problem employee. And sometimes it is necessary to discipline a problem employee. When using discipline as a tool to deal with a problem employee, it is important to use techniques that are fair and equitable for all employees and to use discipline techniques to reinforce the proper performance based on established standards.

You cannot make up the techniques, punishments, and rules as you go along. I am a lifelong fan of NASCAR racing — one of the products of growing up in North Carolina. However, during the past several years, it appears that fines and punishments for not adhering to the standards are done on-the-fly and the rules are changed to fit the whims of NASCAR leadership. You cannot do this in your supply chain operations. Standards must be written and established before you can use disciplinary tactics to punish an employee.

Another use of discipline is to actually improve performance. If standards are developed and published — as well as communicated to the employees—and the standards are not being met, discipline measures may be necessary to improve performance. In this application it is important that the discipline is matched to the shortfalls. This form of discipline may actually include additional training on the equipment, procedures, and techniques to assist the employee in improving his/her performance. Kellogg Bakeries in Charlotte, North Carolina, implemented a Safety School on Saturday mornings for employees who had safety accidents on the job. This school was not designed and implemented as a punishment but rather as a way to provide employees with additional safety training on the job site. The result was a greater appreciation for plant safety and a reduction in safety accidents at the plant.

Decisiveness

A leader's job does not end when the decisions are made.[5]

Decision making is only the beginning of a leader's responsibilities. A leader uses a sound process to make the decisions and then provides his/her subordinates with the resources to become successful. One way of looking at decision making is to start with the desired outcome and then assemble the facts and assumptions that surround the situation. After reviewing all the facts and assumptions, a set of alternatives must be developed. These alternatives are then analyzed and compared, and the best alternative is then chosen. This decision-making model is effective whether deciding on where to place the new distribution center or where to go on the family vacation.

Summary

Devotion to your job and to your family, *dedication* to excellence, *discipline* in time management and employee performance, and *determination* in setting and achieving goals will lead to success in any endeavor — in your personal life or in your supply chain operations. As a supply chain leader, you must model devotion to the job and devotion to your family; you must model discipline in work ethic, you must model discipline to do was is right; and you must model for your employees to emulate the determination and dedication to setting and achieving meaningful personal and professional goals.

Dedication, devotion, determination, discipline, and decisiveness are the earmarks of leaders in any field. In supply chains, these attributes become even more important. To develop these attributes in your supply chain organizations, you must provide the model for your employees to see, emulate, and follow. These attributes are not something that you can preach but not follow; your employees are looking to you to see if you are just talking about these attributes or if you are actually practicing what you preach.

Notes

1. Powerlifting is a combination of three lifts. In the squat, the lifter removes the weight from the racks, moves backward, gets set, and then squats down to a point where the thighs are below an imaginary parallel line with the floor and returns to an upright position. In the bench press, the lifter takes the bar out of the rack while lying on a bench, lowers the bar to his/her chest, waits for a signal from the head judge, and then returns the bar to arms' length. In the deadlift, the lifter squats down, picks up the bar to a fully erect position, and then lowers the bar to the floor on the command of the head judge.

2. The Velocity Management Program started initially to look at the impacts of customer wait times on the readiness of U.S. Army equipment. The program started with customer wait times because the U.S. Army had more than a decade's worth of data on customer wait times and the components of customer wait time. Eventually, the Velocity Management Program expanded to include methods and procedures to improve the impacts of stockage policies (what was actually stocked at what major wholesale distribution centers), the impacts of stockage and customer wait times on maintenance repair and rebuild activities, and the impact of the maintenance rebuild facilities on customer wait times; and the program looked at the impacts of returns on the financial and distribution systems.

3. The Vice Chief of Staff for the U.S. Army is the Army's equivalent of the Chief Operations Officer for a major corporation; and with more than a million Active and Reserve soldiers plus all of the Department of the Army Civilian employees, the U.S. Army is a major corporation by any standards.

4. Waltrip, Darrell. 2007. In his Christmas letter.

5. Axelrod, Alan. 2006. *Eisenhower on Leadership*, John Wiley & Sons, Jossey-Bass, San Francisco, p. 24.

Chapter 7 Questions

1. What important items do you have on your calendar, and what important activities do you attend for your family?
2. How quickly do you make decisions?
3. Do you set realistic goals and then have the determination to continue on to those goals?
4. What goals do you set for your employees, and how do you communicate those goals?

Chapter 8

E4
Equality, Example, Expectations, Enthusiasm

Equality, example, expectations, and enthusiasm have an impact on whether or not an individual is willing to follow you in supply chain operations or in any other endeavor in which you may be involved. The concept of equality is stated in the *United States Declaration of Independence* — "We hold these truths to be self evident that all men are created equal, …"

Equality and Supply Chain Leadership

How do you treat your employees? How do you treat your contractors? Several years ago, my favorite comic strip, *Dilbert*, had the pointed-haired boss giving out shirts to boost employee morale. By the end of the strip, the boss declared none for temporary employees, none for contractors, and none for a few other categories. How many of us use temporary employees in our operations? I would venture a guess that the majority of supply chain operations use temporary employees at one time or another.

We established the Theater Distribution Center in Kuwait using two shifts a day of temporary employees. These "temp hires" volunteered to work twelve-hour shifts in extreme desert conditions. Okay, they did not really volunteer. What they did do was arrive in Kuwait before their equipment arrived and therefore had no real mission. And because they had no mission, they happened to be in the wrong

Figure 8.1   Theater Distribution Center, March 8, 2003.

place at the wrong time and "volunteered" to work for us. Unfortunately, we had a new crew of "temp hires" every twelve hours. Without the use of "temp hires," we would never have been able to go from Figure 8.1 on March 8, 2003, to the cleared yard depicted in Figure 8.2 just 10 days later. I would venture a guess that many of us throughout the supply chain world would not be able to achieve our levels of productivity and excellence without the use of temporary hires. The only full-time employees at the Theater Distribution Center other than my small staff comprised a total of four Bangladeshi forklift drivers complete with four indoor standard warehouse forklifts for use outside in the dirt.

The question based on the *Dilbert* comic strip is: Do we treat temporary hires differently than full-time employees? What about contractors? How many of us depend on contractors to get products through the supply chain? Again, I would venture a guess that all of us do. Do you treat contractors differently? I recently worked in an organization in support of the Department of the Army that was predominantly a contractor operation. In fact, more than two-thirds of the employees were contract employees. However, it was very evident to all the contractors that they were considered second-class citizens in the organizations. The majority of the training products produced by this organization were prepared and trained by the contractor employees. When it was time for praise for the products, the contractors were left out of the ceremonies. When a function was held during normal work hours, the contract employees were told that they could attend if they were not on

Figure 8.2 Theater Distribution Center, March 18, 2003.

the clock — even though the function may very well have been official in nature. Do you treat your contract supply chain employees differently?

Do you give your contract employees the same safety equipment that you provide the company employees? When I was in Kuwait, the most helpless that I felt was on the first day of hostilities. On that day we endured eight SCUD missile warnings, two of which resulted in Patriot missiles fired to destroy the SCUD. At that point in time, we still assumed that the Iraqis had chemical weapons and had to assume that any SCUD fired could be a chemical attack. After the first alert and after putting on my chemical suit and protective mask, I looked over to check on my workers and realized that my Bangladeshi contractors did not have masks or chemical suits. All the U.S.-based contractors and civilian personnel were provided with chemical suits and protective masks. What made the Bangladeshi contractors different? Was there a different standard for different contractors? Where was the "all men are created equal"?

The next morning, my four Bangladeshis were not at the Theater Distribution Center. Because some of the local contractors did not show up, I was not concerned at first. About lunchtime, my Bangladeshis showed up. The "leader" of the group came to me and said, "My friend, we would have come earlier but we did not have masks." We immediately found masks for them, which they kept on their forklifts during the day and took home with them at night. By treating them as equals and showing concern for them, I not only crossed the communication gap with these four workers, but I also discovered the hardest-working forklift drivers in all of Kuwait. They became the first employees to arrive every morning and were the last to leave each night. When I made my final visit to the Distribution Center prior to leaving Kuwait, my Bangladeshi friends all came to thank me and say good-bye.

Unfortunately, this story on equality does not end there. When the designated permanent full-time unit arrived to work at the Theater Distribution Center, they were led by an African-American lieutenant. This lieutenant came to me after a couple of days and said that he could not work with "those workers over there" while pointing at the Bangladeshis. This shocked a few of us. After a little discussion about discrimination and the history of African-Americans in the Negro Baseball Leagues and the hardships they faced while playing for the love of the game and being denied access to Major League Baseball, this lieutenant came around to the right way of thinking.

What about how you treat employees when a company merges with another company? We touched on this subject earlier but it also fits into the equality equation of leadership modeling. The Romans understood thousands of years ago that people who feel like they belong will be more loyal than those who feel like outsiders. The Romans made the citizens of their conquered cities Roman Citizens. This may have been the secret to the longevity of the Roman Empire and its concepts — look at the impact on our calendars, measuring systems, government, and roads.

Maybe we should go back to history class and study the Romans to learn how to treat employees when companies merge. In previous discussions we looked at a couple of companies that merged but continued treating the employees as different organizations. This not only impacts employee loyalty, but also reflects on the leadership modeled by the corporations.

What about other prejudices that impact the attitudes of your employees and therefore the attitude that your employees show your customers and consequently impact your supply chain? How do you treat your employees? Here is another example of treating employees with prejudice. In one Department of the Army organization that I did some work for, I observed that several of the Department of the Army civilians[1] appeared to have a disdain for the contract employees, who made up a large part of the organization. The more I observed this phenomenon, the more I wondered what was driving this attitude. After asking around and even confronting the personnel directly, I discovered that the reason for this disdain and apparent discrimination was based on the fact that the Department of the Army civilians had retired from the Army at lower ranks than the contractors. For whatever reason, this led to professional jealousy that was demonstrated through the lack of professional respect for the contract personnel.

Do you treat all employees the same? Here is another example of not treating all employees the same. In 2002, the U.S. Army National Training Center changed its primary supply chain contractor. This announcement was made in November 2001 with a changeover date of February 1, 2002. All employees were hired in November 2001, and those who were hired were measured for new uniforms. On February 1, 2002, there was a formal ceremony welcoming the new company to the National Training Center family. The employees were all standing on the parade field for the ceremony. What was apparent was that all the employees did not have their new uniforms. All of the management team had new uniforms with the name of the new company on their shirts. Some of the employees had new uniforms; some of the employees had partial uniforms; some of the employees were wearing the uniform of the old employer with the patch with the name of the old company removed; and a couple of the employees who needed XL, XXL, and XXXL uniform sizes ended up waiting several more months for their uniforms. The fact that the managers all had new uniforms was not lost on the employees.

Modeling Supply Chain Examples

Everyone can be used as an example — some as good examples and some as bad examples. How then do you model the right examples for your employees? Every action every day must convey your commitment to both the supply chain employees and the supply chain customers. Everyone in your supply chain is watching you as a leader to see how you respond, how you act, and how you react as the example of how they should act or not act. The last thing you want as a leader is for your employees and supply chain partners to use you as the example of how not to act, respond, or react. The whole premise of this book is to provide you with a guide on what right looks like as you benchmark and model your supply chain leadership.

The choice of whether or not you are modeling the right example is up to you. What legacy do you want to leave behind — an example of how not to respond, act, talk, or react or a lasting example of how to act and how to treat employees, customers, and superiors?

Expectations and Supply Chain Leadership

We all have expectations of leaders and employees. How do you convey those expectations to your employees? How do your customers convey their expectations to you? Earlier in Chapter 7 we discussed setting goals and modeling the determination to achieve those goals.

What expectations are you conveying to your employees with these goals? Are you setting goals for your supply chain that are designed to make yourself look good? Are you measuring your employees against goals that are unrealistic?

I have seen companies that set unrealistic goals for themselves, not realizing that establishing expectations that are unrealistic is a good way to de-motivate your employees.

The expectations that you are looking for in performance from your supply chain should be set from the customer perspective and measured from the customer perspective. When we discussed Sun Tzu in Chapter 2, we looked at the concept of knowing yourself and knowing your competition, as well as knowing your customers. In the discussion on employees, we discussed the concept of knowing your employees. Both of these concepts are important when establishing the expectations for your employees and your supply chain as a whole.

As a leader, you must understand that a one-size-fits-all set of expectations from your employees is not going to work. Employees produce at different rates and therefore expectations must be different.

One company that I visited a couple of times on the West Coast established expectations (goals) for items picked per hour and tied employee pay raises to this metric. What I noticed in watching the distribution center employees was very revealing. I spent a few hours watching the picking operations in one particular zone of the distribution center. There was something wrong but I could not put my finger on it.

There was a large scoreboard that showed the number of items that each employee picked per hour, and the department manager updated the board every hour. I noticed that some of the employees grabbed the first pick ticket in the queue and went to work on picking that item. But some of the employees in this particular department were flipping through the pick tickets, grabbing one from the interior of the queue and then picking the items. This appeared a bit strange but I was not sure why.

So I watched a while longer and talked to the manager about the scoreboard. I thought the scoreboard was a great idea because it showed the goal or expectation of the employees and then showed by hour how well each employee was doing against the stated expectation. Then I started noticing that the employees at the top of the scoreboard were the ones who were flipping through the pick tickets.

I started to suspect that there was a correlation between flipping through the pick tickets and the leader board. It was not until I started asking questions that I was able to put it all together. There was a correlation between the stated expectations of the employees, the ability to meet those expectations, and the promotions/pay raises. More questions revealed that the veteran distribution center employees had cracked the code that linked the expectations and the pay raises. They were flipping through the pick tickets to find the easier picks and the SKUs that were closest to the shipping dock to reduce their travel times and therefore improve their number of picks per hour. You must be careful how you communicate your expectations and how you measure you employees against those expectations. They will modify their behavior to meet those expectations.

What do your employees expect of you as a leader? Most employees in the supply chain business are still motivated by Abraham Maslow's "hierarchy of needs" but that does not reduce the need to provide a safe, secure workplace free from any type of harassment. Employees in all businesses expect leadership that is concerned about the employee, leaders who are interested in the employee, and we all want leaders who are visible.

What about your supply chain customers? What do they expect of your supply chain? If all customers are like me, what they want is nothing less than perfect order fulfillment. Realizing that perfect order fulfillment is not always possible (one company actually told me that calculating perfect order fulfillment was "just too hard to do"), what customers want is on-time delivery[2] and communications about their orders.

Enthusiasm and Supply Chain Leadership

Just what is *enthusiasm*? The *Merriam-Webster Online Dictionary* provides a good definition of enthusiasm: "strong excitement of feeling; something inspiring zeal or fervor."[3] Is it important for a leader to be enthusiastic? Absolutely! How can a leader provide purpose, direction, or motivation to the people that he/she leads if the leader does not have enthusiasm and passion for the job he/she is doing? Why would you follow a person who did not believe in what he/she is doing and was just going through the motions? Are there really any benefits to enthusiasm? A leader who can ignite or fan the flames of employee enthusiasm can get his/her employees to do almost anything. The true benefit to the leader with motivation and enthusiasm is an organization that is filled with people with the fires of enthusiasm. This organization will never have to worry about motivational dysfunction.

Can you fake enthusiasm? You can fake anything but your employees will know if you are faking enthusiasm. I once had a commander whose driver faked enthusiasm. When I was stationed in Germany in the mid-1990s, my commander's driver always answered every question about anything with, "Excellent." This particular driver was a great kid who was not really happy in the U.S. Army so his answer to anything was a faked enthusiasm and the reply of "Excellent." A large majority of the people who he came in contact with thought he was a "gung-ho" soldier because of his apparent attitude. Those of us who spent time with him knew that he was faking the enthusiasm to stay in the cushion job of driving rather than actually have to work as a mechanic.

More than fifty years ago, Dr. Norman Vincent Peale penned the classic *Enthusiasm Makes the Difference.* He was on the mark when he wrote this book. Do you need enthusiasm in your supply chain to be successful? No. I have worked with several companies that were just going through the motions with little or no enthusiasm. Of course they also wondered why they were losing customers and had low employee retention rates.

I recently came across a high school athletic director who had no enthusiasm for the job whatsoever. He carried a decades-old grudge that he did not get a college scholarship out of high school so he went out of his way to not assist his students in getting scholarships. Rather than try to motivate the athletes at his high school, he was content to stay within the guidelines and enjoy his status rather than help the students, and he actually yelled at some of the athletes for showing initiative.

During a visit with him, I asked, "Would you rather have a student who is enthusiastic and takes actions on his own initiative and who you may have to reel back in, or would you rather have students who do not act at all?" As a person in a position to mold young athletes and students, his response shocked me. He said he would rather have students who wait to be told what to do. What would you prefer as a supply chain leader — employees who show enthusiasm and initiative, or workers who sit around waiting for someone to tell them what to do? I prefer the employee who I may have to reel in occasionally to the one I have to constantly prod and guide.

As a leader, why would you want to dampen the enthusiasm of your employees? Or, as a coach, why would you ever want to dampen the enthusiasm of your players? I cannot conceive of a time that this could ever be productive for a team or an organization. And yet, it happens every single day in supply chain teams and athletic teams. Players/employees arrive with a high level of enthusiasm (for the most part) and their manager/coach/leader publicly criticizes them or ridicules them in front of their peers — the result: dampened enthusiasm, frustration, even quitting the program or the company. Another outcome is motivational dysfunction — the "why try?" attitude.

Enthusiasm once lost is hard to regain without a change in attitude or change in leadership. The fires of enthusiasm can ignite the enthusiasm in others; but when the fire is extinguished by untrained or worse, unthinking, managers/leaders/coaches, it is very hard to reignite the fire — burnt embers do not easily reignite.

An employee with too much enthusiasm or initiative can be reeled back in a little and still have a positive impact on the organization. However, an employee whose enthusiasm fire has been extinguished may be hard to remotivate or reignite. When the enthusiasm fire has been extinguished, not only does motivation dysfunction set in with that employee, but it also starts to spread to other employees.

In professional sports, the solution appears to be easy. The solutions are to fire the coach/manager and bring in a new staff, or to trade the newly unmotivated player to a new team. In business, especially our business — the people business of supply chains — the answer may be as simple as providing leadership coaching and training to those employees we place in leadership positions.

Just as being a teacher does not automatically make a person a leader or a coach, being a good manager does not make a person a good leader without additional training and education. Shortfalls in leadership training can lead to accidental or intentional extinguishing of employee/athlete enthusiasm and the planting and fertilizing of the seeds of motivational dysfunction.

Dr. Peale was correct: enthusiasm does make a difference and, as a leader, not only do you have to model enthusiasm, because it is contagious, but you also have to work hard not to extinguish the fires of enthusiasm in your employees. The choice is yours: model and encourage enthusiasm, or extinguish the fires of enthusiasm and thereby encourage motivational dysfunction.

Enthusiasm is linked to your passion for what you are doing, but not to your ego. Although we discuss passion in more detail in Chapter 13, it is important to point out here the link between your passion for what you are doing and your enthusiasm. A person cast into a position simply for money will not have the same passion and enthusiasm as the person doing the job because he/she believes in the mission and really wants to be there. The unfortunate result of being in a job that you do not want to be in is a spreading of motivational dysfunction through your organization or team and taking the fun and enthusiasm out of the program. This happens all too often in life; if you are not having fun because you do not want to be in a specific job or leadership position, leave before you pass that lack of enthusiasm to the entire organization. Lead from the front — model enthusiasm and fan the fires of enthusiasm of those you lead and coach.

I recently watched two different coaches extinguish the fires of enthusiasm in their athletes. One coach was untrained for the job. To make sure her favorite athletes were taken care of, she created an atmosphere of unequal treatment and public ridicule for those who tried to improve the team's attitudes. When she had problems, instead of dealing with them, she went to the school activities director to discipline the athletes who had enthusiasm. The activities director was a tenured employee with no coaching skills. I used the past tense "had" because they successfully extinguished the fires of enthusiasm. This coach's actions had other impacts. Her daughter was a member of this team and became ostracized by the other team members because they perceived her as a spy for her mother the coach, and the unequal treatment also turned lifelong friends against each other.

The other coach allowed his ego to get in the way. His desire for personal glory took the wrong road — as it will usually do. In one year he went from running a successful, highly motivated, enthusiastic program with team-building exercises at the end of every practice to telling his athletes, "I want a World's Bid this year and if you do not like it, there's the door." Only a third of his athletes stayed with the program from his National Championship team from the year before. The fires of enthusiasm were extinguished by the coach's ego and with the extinguished enthusiasm, the fun of the program left as well.

When you fan the fires of enthusiasm, you will be amazed at what your team, your organization, or your department will achieve. You will be amazed at the number of people who want to join your team or department because of the fires of enthusiasm. What team do you want to be on? The team that suffers from motivational dysfunction because of a lack of leader and employee enthusiasm, or the team that everyone wants to join because of the rising flames of enthusiasm?

As a leader, you can create such a situation by modeling the enthusiastic behavior for others to emulate. Enthusiasm is contagious. But remember that you cannot fake enthusiasm; you cannot fake equality. Genuine and real are the benchmarks of enthusiasm and equality, and are the examples that you want to model for your supply chain team.

Example of How to Kill Enthusiasm and Initiative

As a coach or leader, would you ever go out of your way to make a player or team member look bad? I recently observed a coach who would do the opposite of whatever her most experienced captain would suggest. I could not ascertain if it was incompetence, inexperience, ego, or just personal jealousy. Whatever the reason, she would not take suggestions from this particular team captain and threatened the captain with removal from the team if the captain made any more suggestions.

This particular coach was inexperienced and new to the sport. Her "qualifications" were threefold:

1. She was already a teacher in this school system.
2. She was a good friend of the athletic director (not necessarily a qualification for coaching).
3. She had one year of experience, that of watching her daughter play this particular sport.

You will notice that there is no coaching experience listed as one of her qualifications.

To make up for her lack of experience, she attended a 2-day seminar and was convinced that the seminar made her sufficiently qualified to be the head coach. And because she "knew everything necessary to coach" (her words), any suggestions from the most experienced captain were taken as a challenge to her abilities and reported as "disrespectful comments" to her "friend" the athletic director.

This situation became so bad that the other girls on the team would get the ideas from the experienced captain and pass them to the coach as theirs to get them implemented. Not a good situation for any of the team members, but it did get the right things done for the team.

The final result of this situation was a total loss of respect for the coach by the players (except her favorites); ostracizing of her daughter by the other team members because they were afraid she was a spy for her mother the coach; and a total loss of enthusiasm by the most experienced athletes, resulting in a dysfunctional team.

How does this translate to your supply chain? Building an enthusiastic, functional supply chain team is just as important as building an enthusiastic, cohesive athletic team.

Are you going out of your way to build an enthusiastic team for your supply chain? Or, are you like the coach in this example and letting your ego or inexperience get in the way of winning and success? Your employees are your legacy. Are you leaving an enthusiastic legacy or a dysfunctional one? The choice is yours.

The coach, like every other leader, provides purpose, direction and motivation to his/her team.

Notes

1. In Department of the Army organizations that are not combat units, there are three components of the workforce. Obviously there are soldiers (usually there are soldiers in charge; however, there are organizations that are "led" by Department of the Army civilians); there are Department of the Army civilian civil servants; and there are contract employees (usually retired or former military members).
2. In this case, on-time delivery is delivering the product at the time it is promised when the order is placed. This may very well be your competitive edge over your competition.
3. http://www.merriam-webster.com/dictionary/enthusiasm. Accessed October 5, 2008.

Chapter 8 Questions

1. What can you do to fan the flames of enthusiasm for your supply chain team?
2. Are you intentionally or unintentionally extinguishing the flames of enthusiasm?
3. Are you treating all your employees the same?
4. Is there an illusion of partiality in your organization? Why? What is causing the perception or illusion?

Chapter 9

R3
Respect, Responsibility, Reliability

> You have to give respect to get respect.
> —Joe Torre, Manager, Los Angeles Dodgers

Respect, Responsibility, and Reliability are critical to supply chains, critical to supply chain leaders, and critical to leaders at all levels in any organization.

The 1990s brought us the concept of being disrespected. Although it entered the lexicon via slang, the idea of respect for employees, customers, and leaders is not a new concept. Unfortunately for people at all levels of the organization, the thought process is that respect is deserved without giving respect to others. Respect as a leadership attribute is critical to success in modeling leadership for subordinates to emulate but has to be given in order to be received.

Responsibility is both a personal and a supply chain attribute. All leaders have to be responsible for their actions and the actions of the people working for them. At the same time, your supply chain has to be responsible to your customers. Modeling responsibility by leaders will enable employees to model the behavior that will improve the success of the employees as well as the organization.

Your personal reliability is tied to the attribute of honesty and integrity and, like responsibility, is a personal and organizational requirement for successful supply chain operations. Modeling personal reliability may be as simple as doing what you say you are going to do, when you are supposed to do it, and doing it in an ethical manner.

Respect

Respect is something that we all deal with every day, and not just in our supply chains. Unfortunately, we also have to deal with a lack of respect or disrespect almost every day. What is respect, and what does it have to do with supply chain leaders?

Teaching respect for others can sometimes backfire. The U.S. Army discovered this in the mid-1990s when a full-court press was initiated against sexual harassment after a series of serious incidents at Aberdeen Proving Ground, Maryland. Some soldiers found out that the quickest way to get someone in trouble was to claim sexual harassment and, not only would it get the person or persons in trouble, but those accused were automatically guilty until proven innocent.

A similar incident surfaced in a local high school. The high school implemented a "consideration for others" program to teach tolerance and reduce perceived discrimination and bullying. Supposedly this was a "reaction to the massacre/retaliation murders at Columbine." However, it was implemented nine years after the Columbine murders. Anyway, a student was prompted by a "coach" to write statements against another student. The result was that the student who was supposedly making "racial statements" was called into the activities director's office, counseled, and told that this would not be tolerated — all without ever hearing the other side of the story or asking if it was, in fact, true. This worked so well that the student tried it again — this time based on hearsay — and again the other student was called into the office and yelled at, only to find out that the student who reported the incident had the facts wrong and misunderstood the gossip on which she based the statement.

All people deserve respect until proven otherwise. But once you have acted in a manner to violate that trust, you should not expect to automatically get respect just because of your position in the organization.

The concept of respect is tied to the leadership value of attitude. Your attitudes as a leader may very well lead to gaining or losing the respect of your supply chain partners as well as your customers and your employees. As a supply chain leader, you have to earn the respect of supply chain partners and customers on a daily basis. It must be earned — and earned every single day. There is a direct correlation between your supply chain quality and the respect that you have from your partners and customers.

As a coach, teacher, or mentor, you have to earn the respect of your students or athletes every single day. It does not come with the title of "coach" or "teacher." The quality of your team is a direct reflection of your coaching ability and the amount of respect that you show your athletes. In the supply chain, the quality of your supply chain team is a direct reflection of your abilities as a supply chain coach and the respect you show your supply chain team.

Are you modeling leadership in such a way that you are "respect worthy?" What is *respect worthiness*? It is modeling a behavior and attitude toward your employees or team members that constantly earn your respect and modeling behavior that shows respect for others regardless of how they treat you. Showing a form of reverence for all employees or team members is one way of showing respect for your team members. Many team members show a reverence and respect for their coaches when the coaches have a habit of respecting their team members. Are you modeling the form of respect that deserves reverence? Or are you doing as I observed one coach doing recently by modeling a form of disrespect that had team members turning against each other?

Many theorists have written about the psychological aspects of respect. I promise we will not go there in this discussion. However, some of those theorists have identified institutional respect as an aspect of the respect spectrum. Many institutions are accorded this level of respect. The U.S. Army is one of the most respected institutions in polls every year. I recently had the honor of touring the Pacific Air Forces Headquarters at Hickam Air Force Base, Hawaii. The headquarters building was a barracks along the flight line on December 7, 1941. Like the Arizona Memorial a short distance away, these locations have institutional respect because of their historical significance to the freedom of the United States. Churches, because of their importance in the lives of so many people, are afforded institutional respect.

What about your supply chain? Is your supply chain afforded institutional respect based on the performance of your supply chain? Do people have institutional respect and reverence for the links in your supply chain? Why or why not? If you are modeling the proper behavior and measuring the proper performance of your supply chain, you should be able to achieve an institutional respect for the entire supply chain.

As you model respect for your employees and supply chain partners, does your moral behavior influence the respect shown for you and your supply chain? If you act in an unethical or immoral way, does that impact the respect you receive from your employees and customers? If you violate your integrity, does that impact the respect you are shown? Sun Tzu tells us that if the leadership is weak and immoral, the company will be weak. Consequently, if the leadership of a supply chain organization is weak and acts immorally, the supply chain organization will be weak and will never receive the respect that is possible from strong leaders modeling the proper behavior for their employees to emulate.

Is there a moral obligation to show respect for your employees and supply chain partners? I would argue that there is, indeed, such a moral obligation, even when they display acts that do not show respect for you. Early teachings tell us to turn the other cheek. The same is true for showing respect in the supply chain. As hard as it is at times, our customers need to be shown respect even when they act in a manner that does not normally deserve respect. Although respect must be earned, the greatest pleasures in life sometimes come from showering kindness and respect on someone who does not deserve it.

Never intentionally go out of your way to disrespect someone. Here is an example of a teacher who wanted to be a coach. This particular coach needed something that one of the team captains had at home. However, the captain went home early from school sick and was not able to attend practice. The captain contacted another team member to have that team member come by her house and pick up the necessary materials. The coach not only did not thank the captain for arranging to get the materials to the practice, but also had the lack of respect for herself and the athlete to ask the other team member, "Did she look sick?" And she asked this in front of all the other athletes. The coach was not smart enough to realize that she made herself look small and disrespectful by modeling a behavior that the other athletes would not want to emulate.

The *Stanford Encyclopedia of Philosophy*[1] asks the question, "Must persons always be respected?" After repeated actions of disrespecting others, some people such as this particular coach may prove that at some point, respect is not a given and if you continually do not show respect, you will not continually get respect. The same is true for your supply chain; if you continually model a behavior of disrespect, you cannot expect customers, employees, and partners to show respect for you.

Another aspect of respect that is important in the supply chain arena is the concept of *respect for the environment*. We are stewards of the Earth. What we do, according to Native American lore, impacts seven generations. As supply chain leaders, we must model a behavior of respect for the environment for our employees to emulate. Our European partners are way ahead of us.

My last assignment to Germany was from 1992 to 1995. During that time, the environmentally friendly laws were way ahead of what I returned to even in environmentally friendly California. Items that were mandatory recycling items in Europe were and are still thrown away in the United States. In graduate school in the mid-1980s, I had to search all over Brevard County, Florida, to find someone who would recycle the newspapers that I was required to read every day to stay current on business affairs. Packaging reductions are starting to gain footholds in the United States; however, there are still wastes and excesses in packaging that could be reduced to help supply chains earn greater respect by showing a greater respect for the environment.

The *Stanford Encyclopedia of Philosophy*[1] states, "Respect is not something individuals have to earn or might fail to earn, but something they are owed …." However, in our business, the supply chain business, as in every other walk of life, respect must be earned. Although there is the reverential respect we spoke of earlier, even that respect can be lost when leaders act in a manner that does not show good judgment or respect for others. Joe Torre, manager of the Los Angeles Dodgers Major League baseball team, said it best: "You have to give respect to get respect." No one in any position, especially a leadership position, is above this basic law of respect.

> The willingness to accept responsibility for one's own life is the source from which self-respect springs.
> —Joan Didion

Self-Respect

The *Stanford Encyclopedia of Philosophy* states that "The value of self-respect may be something we can take for granted, or we may discover how very important it is when our self-respect is threatened, or we lose it and have to work to regain it."[1]

Another key aspect of respect that is relevant to this study of leadership is the concept of *self-respect*. If you do not respect yourself, how can you expect others to respect you? And if you do not respect yourself, it becomes very hard for you to show respect for others. When you apply the law of respect discussed above, you cannot give respect if you do not have self-respect and therefore cannot get respect from others. To have self-respect as a leader, you must be convinced that what you are doing is both fulfilling and meaningful. If you do not enjoy your job, you will probably incur a lack of self-respect. If you are convinced that your employer is not utilizing you properly, you will probably incur a lack of self-respect. When this happens, your performance declines and you start suffering motivational dysfunction, which can infect your entire organization.

As a supply chain leader, your employees are looking to you. What will they see? Self-respect radiating out, or something less? What are you modeling for them to emulate? However, there is a big difference between self-respect and self-worship. Do not let your ego get in the way of leading your supply chain. Self-worship is when you start to believe that you are better than others around you and act in such a manner that not only does not show respect for your colleagues, but actually shows an attitude of disrespect for them because of the ego getting in the way.

Growing up, I found a poem in my Dad's desk drawer entitled "The Man in the Glass." I have kept a copy of the poem in my notebook and on my filing cabinet in my office since I left home for the U.S. Army more than thirty years ago. The moral of the poem is that you can fool the whole world but if you do not have the respect of the person looking back at you in the mirror, you have not been successful. You have to have a healthy self-respect to be successful in any business — and the supply chain industry is no different. You have to give respect to yourself as well as others in order to be respected.

Figure 9.1 provides a short questionnaire from www.goodcharacter.com; it is a quick check on being respectful toward others. It is from a teaching guide for high school students but seems appropriate for supply chain leaders as well.

When you lose the respect of your supply chain customer, can you get it back? The short answer is yes — but you will have to work very hard every day to get it back. The old saying in the U.S. military is that "Aw, crap can wipe out a thousand 'atta-boys'." The same is true in the supply chain industry. One upset customer wipes out all the good work that you have done over the years to gain and keep their trust and respect. Now the challenge for the leader is to model the right behavior for all of his/her supply chain to see how to work to get back that respect.

It is possible but it is a challenge. Johnson & Johnson/McNeil Labs learned this the hard way in the mid-1980s when tainted/tampered Tylenol was discovered.

Are You a Respectful Person?
True/False

- I treat other people the way I want to be treated.
- I am considerate of other people.
- I treat people with civility, courtesy, and dignity.
- I accept personal differences.
- I work to solve problems without violence.
- I never intentionally ridicule, embarrass, or hurt other people.

From: *Teaching Guide: Respect,* **www.goodcharacter.com,** used with permission

---

Figure 9.1 Questionnaire. (From www.GoodCharacter.com. Copyright Elkind+ Sweet Communications, Inc. / Livewire Media. Reprinted with permission.)

Johnson & Johnson worked quickly to get the tainted tablets off the shelf and replace the bottles with tamper-resistant packages that have now become a standard in almost every consumer product industry.

The same is true when you lose the respect of your employees. Earning back their respect is a difficult chore that must be accomplished successfully if you are going to remain their leader.

Responsibility

> If you want children to keep their feet on the ground, put some responsibility on their shoulders.
> —Abigail Van Buren (Dear Abby)

The same is true for employees. If you want to ground them in the reality of the workplace, give them some responsibility. Responsibility was not part of the original design of the House of Leadership and the cure for motivational dysfunction. After a presentation on "Getting Back to the Basics: Leadership" in South Africa in 2007, an attendee at the conference approached me and said, "I like your model but I think you should consider adding responsibility to the equation." She was correct and, thanks to that suggestion, responsibility as a leadership enabler was added to the modeling process.

What is *responsibility*? Can you delegate responsibility? *Webster's Online Dictionary* defines responsibility as "the quality or state of being responsible."[2] Like so many definitions, this does not really clarify the meaning of the concept of responsibility. Another online dictionary provides a little better definition: "a particular burden of obligation upon one who is responsible."[3] I am not sure that responsibility is always a burden but it is a full-time job for leaders. Taking responsibility for your own actions all the time may be a dying art in today's society.

As this chapter is being written, the U.S. Congress is debating how to bail out homeowners who knowingly bought homes well beyond their means and ability to pay. At the same time, Congress is trying to determine how to bail out the banks that knowingly made the bad loans. The result will be that people who bought homes that were within their ability to repay (responsibility) will probably end up paying extra taxes for those who acted without responsibility. Amazingly, it seems that "it was not my responsibility to buy a smaller home that I could afford when the rates went up." And apparently it was not the responsibility of the banks and mortgage firms to really qualify applicants for homes that they could afford.

A similar problem exists in the state of Florida with homeowners insurance. Every home that I have ever purchased required that I maintain adequate insurance to cover the house and anyone on my property against damages, hurricanes, wind, etc. However, apparently as part of all the mortgage problems, a large number of individuals were allowed to have homes without insurance. They did not take the responsibility for their homes. The result is that homeowners who did indeed insure their homes against peril are now paying more insurance to cover those who did not insure their homes.

What does this have to do with supply chain leadership? Everything! Responsibility represents those things that leaders and employees are expected to do. In the supply chain we are responsible for getting the products from the source to the ultimate consumer of the products. Not only that, but we are also responsible for ensuring that the right product arrives in the right condition, at the right time, and in the right quantity. We must take responsibility for our supply chain actions. We are responsible to our customers for the actions we take in providing and delivering products.

Not long ago, a supply chain company had a delivery of products (approximately $3 million worth of electronics). The delivery truck backed up to the dock doors, and the driver dropped the trailer and departed. Less than thirty minutes later, another tractor hooked up to the trailer and took off. Who was responsible for the loss? Was it the delivery company? Perhaps the receiving company that had not even opened the dock door or the trailer? The last time I talked to the transportation director for this company, the insurance companies, the investigating police department, and the two companies were still trying to determine who was responsible for the loss.

In 2007 and 2008 toys were discovered with pellets that produced drug-like effects and toys for little children that were painted with lead-based paint — who would have thought that a small children would put a toys in their mouths? That has never happened before. Not! So, who was responsible for the problems? In the supply chain world, we are responsible for the actions of our supply chain.

Personal Responsibility

Another aspect of responsibility is the personal responsibility that comes with the position or assignment. You can delegate authority but you cannot delegate your responsibility. In the U.S. Army, I could delegate someone to sign certain actions for me; but if the actions were wrong, I was still responsible for the consequences. The same is true in your operations; you can delegate a subordinate manager to take care of certain tasks in your distribution center but that does not relieve you of the responsibility for ensuring that the actions taken are completed and correct.

At the same time, as you model leadership for your supply chain, it is important to remember that you should give subordinates the authority to act in the best nature for the supply chain or the company. And then if something does not go according to the plan, you must still accept the responsibility for the actions as if you were the one acting. The bottom line is that you are responsible for the actions or inactions of those working for you. When your employees see that you are willing to allow them to try something different and then provide the top cover of accepting the responsibility for what goes wrong while enabling the employees to get the praise for the good things, you will earn their respect and loyalty.

Reliability

Teaching operations management includes teaching systems reliability, which is something many learn in high school physics classes. The topics include systems in parallel and systems in sequence, as well as the calculations for determining system reliability. The study includes discussions of Mean Time Between Failures and Mean Time To Repair. System reliability can be defined by the equation in Figure 9.2.

These topics also impact discussions of supply chain reliability, as well as personal reliability.

As leaders, we are responsible for the reliability of our supply chains. At the same time, we have to model personal reliability for our employees to emulate. The Institute of Electrical and Electronics Engineers defines (IEEE Std 610.12 1990) reliability as "the ability of a system or component to perform its required functions under stated conditions for a specified period of time." What is the ability of your supply chain to perform its required functions under the business climate, business cycles, and business environment over time to support your customers? Six Sigma looks at reliability as close to 99.99999% accurate. Is your supply chain that accurate? Why not? I realize that we do not live in a zero-defects world but our customers certainly expect as close to zero defects as possible.

System Reliability

$$\text{System reliability} = \frac{\text{Mean time between failure}}{\text{Mean time to repair}}$$

Figure 9.2 System reliability.

Perfect Order Fulfillment

Perfect order fulfillment = % on time delivery
×% Right quantity
×% Right condition
×% Right product
×% Right billing/invoice

Ex: If everything is at 99% the perfect order fulfillment %
= 0.99 × 0.99 × 0.99 × 0.99 × 0.99
= 0.95099 = 95% perfect order fulfillment

Figure 9.3 Perfect order fulfillment calculation example.

Is the metric of perfect order fulfillment a measure of your supply chain reliability? It is easy to make a case for that. Perfect order fulfillment is defined as the right product, at the right time, in the right quantity, in the right condition, and the right billing/invoice. Using this as the metric, the calculation would be as depicted in Figure 9.3.

Is 99 percent across the board good enough? And how are you calculating the metric? If you are not calculating any metric from the customer perspective, you are not looking at it from the proper view.

Personal Reliability

Like other metrics for measuring and modeling leadership, personal reliability is an all-or-nothing metric. Would you want to work for someone who was not reliable? No, and neither do your employees. Everyone wants to work for someone who they can trust. When you commit to do something, keep your word and do it. If you do not want to do something, you can say no to the commitment. Your personal reliability is tied to your personal responsibility and your reputation as a leader.

Summary

Your personal respect is tied closely to the amount of reliability and responsibility that you display. Respect is also tied to the concept of taking responsibility for your actions as well as the actions of your employees. Your supply chain respect is also tied to your supply chain reliability. What model are you displaying for your employees to emulate in respect — are you giving respect? Are you responsible for your actions and for those of your employees? Are you and your supply chain reliable? Do you have backup systems in place to ensure your supply chain reliability?

Notes

1. The *Stanford Encyclopedia of Philosophy* is available free at http://plato.stanford.edu/entries. Accessed on September 9, 2008.
2. http://www.merriam-webster.com/dictionary/responsibility. Accessed September 26, 2008.
3. http://dictionary.reference.com/browse/responsibility. Accessed September 26, 2008.

Chapter 9 Questions

1. Do you show respect for others even when they may have taken actions that show that they do not deserve respect?
2. How do you calculate your supply chain system reliability? Who in your organization knows how this is calculated and communicated?
3. Do others perceive you as reliable and responsible?
4. Is your supply chain perceived as reliable and responsible, to include taking care of damages or loss?
5. What actions can you as a new leader take to improve the respect for your supply chain by improving the reliability of the supply chain?

Chapter 10

S2
Self-Development/Employee Development, Serving Leadership

> To rely on rustics and not prepare is the greatest of crimes; to be prepared beforehand for any contingency is the greatest of virtues.
> —Sun Tzu

What Sun Tzu is telling us in twenty-first century supply chains is that as leaders we have to not only develop ourselves, but must also develop our employees. Just because we know how it used to be done — that is, Sun Tzu's "rustics" — does not mean that it is right or that we can rest on our laurels and be content with knowing outdated processes and technologies. To prepare ourselves and to prepare our employees and supply chain team members for any possible contingency or customer demand is "the greatest of virtues" to which Sun Tzu referred. To continue to do things the way we have always done them just because "we've always done it that way" or to not train ourselves and our employees is the twenty-first century crime that Sun Tzu wrote about in 500 BC.

Professional Development

Self-Development

Exactly what is *self-development*? One dictionary defines it as "Development of one's capabilities or potentialities."[1] That really leaves it wide open for interpretation. For our purposes we will define self-development as any action taken by an individual to improve his/her knowledge or skills of his/her profession or any other area that will improve his/her self-confidence or expand his/her knowledge. This can be training — formal or informal — on the job or hands-on. Self-development can also be education to improve knowledge and/or skills.

Self-development is about making a commitment to discovering the skills and education that will help you develop your professional potential. You can improve yourself constantly, or you can stand still and wonder why peers are bypassing you. Self-development, just like corporate improvement programs, does not manifest itself in overnight steps. It takes time and commitment. There are no shortcuts to true professional development.

Self-development can take many forms. APICS (The Association for Operations Management) has a couple of great certification programs. These programs can be completed through classes or through self-study. The most common reason I hear about not completing the program is, "The company does not cover it" or, "I do not have access to the classes." Both of these are excuses, not valid reasons. Individuals who really want to advance in their profession will make the sacrifices and dedicate themselves to achieving a new level of professional excellence and competence.

One of the lessons taught to me early in my military career was that no one had a bigger responsibility for my professional development than I did. The same is true for your professional self-development. No one is going to spoon-feed you what you need. A good mentor will guide your development but even a good mentor cannot force you to develop and advance your skills and professional abilities.

To establish a self-development program, you have to know where you were (where you came from), where you are, and where you are going (where you want to be in your career). To know where you want to be in your professional development, you will need a plan. Otherwise, how will you know when you get there? We discussed setting goals in Chapter 7 and we will look at planning in more detail in Chapter 13.

What areas do you need to focus your personal professional development program on? I recently heard a radio show on the way to the airport where the speaker caught my attention when he said, "The majority of Americans are ignorant." I thought this was a very bold statement and because I had not heard what prompted that statement, I listened more to hear about where we are ignorant. I was surprised to hear that he was speaking at a local university on cultural diversity. He later repeated his statement about ignorance but this time added the caveat that Americans are ignorant about cultural diversity.

Cultural diversity is definitely an area that all leaders need to work on as part of their personal professional development. The biggest complaint about workers in warehousing at the end of World War II was that they did not speak English. One of the biggest complaints in warehousing today is that workers do not speak English. Imagine that in sixty years we as an industry have not progressed very far in our cultural diversity training and understanding. The solution to the language problem is as simple now as it was sixty years ago — learn basic communication skills in another language. When I lived in Germany, the standard line was, "If you speak three languages, you are trilingual; if you speak two languages, you are bilingual; and if you speak one language, you are probably American." And yet we require two years of a foreign language in almost every school district in the country to graduate from high school.

I was in a distribution center in Southern California a few years ago where, not surprisingly, the majority of the workers spoke Spanish — a few spoke English as a second language. However, the problem came from the fact that only one supervisor spoke Spanish. It was like watching an old Kung Fu movie with subtitles. I always feel that I am losing something in the translation when the dialogue lasts for about a minute and the subtitle says, "Hi, how are you?" I made some suggestions to the supervisor and when he translated it to the workers, I was not sure that all of what I said was translated. I speak only a little Spanish but did not hear everything I said when it was translated.

There are a couple of other reasons for having a professional development program to improve cultural awareness. The first is that as our supply chains have become global and expanded, we need to understand the cultures of the workforce that we have working for us in our home country as well as the cultures of the workforces that we have working for us or as part of our supply chain team in other countries in order to understand the motivations of those workers.

The other reason for professional development in cultural diversity for supply chain leaders is that the cultures of different countries dictate how business is conducted and how negotiations are conducted. What is considered standard practices or acceptable business behavior is not necessarily the same from country to country, and supply chain leaders working for multinational countries and conducting sourcing and supply chain functions in other countries need to study and understand the culture of the country in which they doing business.

Negotiations

Supply chain leaders in the twenty-first century need some formal training in negotiations. As leaders climb the corporate ladder and assume more responsibilities in their jobs, they need to have some training in negotiations in order to conduct business in different regions and countries. Just as business practices and mores are different in different countries, so are the negotiating styles. The nexus between supply chain operations, cultural diversity, and understanding different national negotiating styles is where the supply chain leader needs to excel.

Other Key Professional Development Areas

The following is not an all-inclusive list of professional development areas that supply chain leaders need to master in order to excel in this business but does provide a good starting point for your professional development plan and in preparing professional development programs for your employees.

1. *Public speaking.* This one area is still listed in almost every publication as the number-one fear of all people. Supply chain leaders must be able to communicate with their employees, employers, and supply chain partners. As a supply chain leader, you will be called upon to prepare business briefings, updates to the boss, or supply chain professional development programs for your employees. If you cannot communicate your programs, goals, and accomplishments to your employees and employer, you may miss out on the chance to move your supply chain to the next level.

2. *Written communication.* Just as important as speaking skills for leaders is the ability to communicate clearly and concisely in written communications. As I read papers written by some of my graduate students, I start to worry that as a country we are losing the ability to communicate effectively in writing. I once had a boss who made it very clear that you lose credibility when you submit a paper or presentation with grammatical errors. His theory was that the audience gets so wrapped up in looking for more grammatical errors that they lose the points that you are trying to make in the presentation. I once sent a new junior leader to a course on written communication skills to improve her ability to write and communicate effectively. She was a college graduate with a good grade point average (GPA) but her ability to write more than a few sentences was limited. She was initially upset but after the course was very appreciative that someone cared enough to help her improve. As leaders in today's supply chain, our credibility is on the line if we cannot communicate effectively in writing.

3. *How to think.* Knowing how to think is critical. Unfortunately, graduates of many business programs and supply chain professional development programs fall into the trap of being taught what to think and not how to think through problems. There is always more than one way to look at a problem and come up with solutions or recommendations. Many corporate programs run their new leaders through a problem on problem solving that teaches the corporate way as *the* way and thus train their new leaders and employees *what* to think and not *how* to think. As a leader, look at every problem as something new and try not to apply the "old rustics" to the situation.

4. *Hands-on practice.* Leaders need to occasionally get some hands-on practice in the field. You cannot lead from the office; you need to know what is going on in the field. General Gus Pagonis knew this well when he retired from the U.S. Army and took over the supply chain leadership role for Sears. Pagonis made all his supply chain leadership staff go out in the "field" on a regular basis to see what was going on and what the impacts of their corporate decisions had on the rank-and-file worker. Some of the folks went to the distribution centers to work. Pagonis went to one of the local stores to work in the backroom. He was convinced that the lady he worked for thought he was just a retiree trying to make a few extra dollars. Little did

she know that the Vice President of Sears was working in her storeroom. Spending time doing the work occasionally also allows leaders to see the conditions that the workers are experiencing and the impacts on operations that these may have on operations. Driving a forklift in Kuwait allowed me to understand the impacts that heat, rain, and rocket attacks had on the operations at the distribution center and enabled me to explain to my bosses the impacts that these had on our ability to load and unload trucks.

5. *Handling the problem employee.* We discussed this a little in Chapter 7 when discussing discipline. I believe it is important to have some formal instruction in handling the problem employee. My first course in this was in 1984. That class opened my eyes to the problems that not properly documenting problems on the job can have in removing an employee from the workforce. In today's society, improper handling of the problem employee may very well end up in the court system. One local restaurant close to home had an employee who was pregnant. This was not the problem but her pregnancy did cause her to be late a few times and left her unable to work on some other days. Instead of properly handling this situation, the owner of the restaurant fired her for being late. When this case went to court, everybody lost. The owner declared bankruptcy, the employee did not get any money for the mishandled dismissal, and the local restaurant customers lost a good restaurant. Handling the problem employee is much like what the U.S. Army does for problem employees. A paper trail is important. Just as you would want a good paper trail for items moving through your supply chain, you should also have a good paper trail detailing every time an employee is counseled for infractions of company policy. It is easier in the long run to have a class on handling a problem employee — and there is not a one-size-fits-all process — than mishandling the problem employee and spending time in court. I once had an employee who was somewhat of a problem. Every attempt to work with him ended up in threats from him to cause problems through the Equal Opportunity Office (where his wife worked). After not being selected for a promotion because of documented performance and being told "do not consider me for the job because I am the only one who can do what I am doing right now," this employee filed an Equal Opportunity Complaint against me for discrimination. His original complaint was that he was not hired because he was "a white male and over forty." At the time, I was — by my best guess — a white male over forty. That complaint did not hold water and was thrown out by the Equal Opportunity Office at my next higher headquarters level. That led to another formal complaint that he was not selected for promotion because of the fact that he was Southern Baptist and "American Indian." This complaint was finally closed out when it was discovered that my wife had enough Cherokee blood in her to retire to a reservation. He could not back up his claims and the paper trail that had been maintained on his performance enabled dismissal of

the case, but only after a two-year prolonged investigation. Make sure you handle problem employees in the right way; always allow them to maintain their dignity while you maintain a paper audit trail on any conversations about their performance. Do not wait until the annual performance appraisal to let problem employees know that they are having problems. Remember that a problem employee allowed to continue is like a cancer growing in the organization. The problems will continue to grow until you handle the problem employee.

6. *Technical skills.* As supply chains continue to expand and depend on new technologies, supply chain leaders need to know the basics of their systems. This includes knowing the automated systems that drive the supply chain operations as well as the systems in place to measure the supply chain processes. New workers are being taught the systems and machines used to operate the distribution center or plant floor. Leaders need to remain current on those same systems and machines to mentor the employees on the supply chain team and to recognize when a system is not working properly or when workers are not performing properly.

7. *People skills.* We are in the people business; therefore we need skills to motivate employees to want to achieve excellence. All too often we take a good manager and promote him/her to a leadership position, or take a good line worker and promote him/her to a supervisory position, and do not give them additional people and leadership skills — and then we wonder why that person does not perform at the level we thought he/she would. We all need people skills — how to deal with employees, how to talk to employees, and how to motivate employees. It is more difficult to motivate employees who used to be your co-workers. Included in the people skills that leaders must have in today's supply chains is the ability to coach and mentor other employees. Coaching skills are not in-bred skills. Coaching is discussed in more detail in Section III. Supply chain leaders must develop coaching skills and teaching skills as part of their personal professional development plan. Part of this skill set is the ability to instill your passion for supply chains to your employees and how to transfer that passion for supply chains into more productive and passionate employees.

Employee Professional Development

> Remember that good judgment comes from experience and experience comes from bad judgment!

The goal of employee professional development programs is to ensure that our employees do not have to learn from bad judgment but rather from our experience and the proper training and education.

Once a leader has established his/her professional development plan and program, he/she is responsible for developing a professional development program for his/her employees. The same skills that are included in leadership development programs should be part of the employee development programs. In addition, employee programs must include both the basics and the advanced techniques for the functions of the distribution center or supply chain operations.

There is a direct link between employee training programs and employee satisfaction. If you look at the FORTUNE© "Best Places to Work in America," you will see that the companies that consistently rank in the top one hundred companies are the ones with stringent training programs. These programs are important for integrating new employees into the workforce as well as for keeping current employees up to speed on new programs and refreshing their knowledge on the old systems and techniques.

Training leads to competent, confident employees. Confident and competent employees lead to competent, confident companies. Competent, confident, well-trained employees lead to customer satisfaction. There is a cost to training employees but there is a bigger cost to not training employees and losing customers.

Serving Leaders

Much has been written lately about the concept of *serving leadership*. Store bookshelves are full of books on serving leaders. This is not a new concept. Leaders have subscribed to this concept for thousands of years. Serving leaders are those who do not put themselves above their employees. Serving leaders are those who do not ask their employees to do anything that they (the serving leaders) would not do.

Serving leaders go out of their way to assist their employees and they put their employees' career progression above that of their own. Employees who work for serving leaders stay in touch with those leaders as mentors long after they stop working for them.

Without trying to shortchange the concept of a serving leader or discount a number of really good leadership books dealing with the serving leader, it is a simple concept that does not need a full book to explain.

The U.S. Army has a value called "selfless service." Selfless service for the U.S. Army is placing service to the country and the needs of the U.S. Army above personal needs. This is a little selfish from a personal point of view, especially in light of the number of deployments some soldiers have accomplished in the past seven years and knowing that a good number of senior leaders have somehow dodged the deployments or used bogus excuses to prevent deploying. Asking employees to place their jobs ahead of everything else is what leads to employee burnout, an inability to find that personal–professional balance that we discussed earlier, and leads to lower employee retention.

The true meaning of selfless service is the definition of a serving leader. Selfless service is placing the needs of the employee above the needs of the leader. It is placing the career progression of the employee above the career progression of the leader. The guidance of the great Green Bay Packers Coach Vince Lombardi to his team was that their priorities were their relationship with God, their relationships with their families, and then their responsibilities to the Green Bay Packers. That is the guidance of a serving leader. Placing the responsibilities of the players' families above the responsibilities of the job is the mark of a serving leader.

A serving leader is a caring leader — all of the values, attributes, and leadership enablers that we have discussed in this book are the traits of a serving leader. Concern for the employee, balanced with a concern for the corporation, placing the employee above the leader, and working hard to improve the employee and thus improving the company, are the marks of a serving leader.

Chapter Summary

Serving leaders strive to put their employees above their own professional goals and ensure that their subordinates get the proper training that will assist their employees in progressing. The opposite of a serving leader is one who puts his/her ego and career progression above that of their employees. A non-serving leader could care less if his/her employees get promoted or even stay with the company. A serving leader has a very high employee retention rate and a highly motivated workforce.

Leaders at all levels are responsible for their own professional development and the professional development of their employees.

Notes

1. Self-development. http://www.ask.com/web?q=Definition+of+Self+Development& qsrc=6&o=10616&l=dir. Accessed October 13, 2008.

Chapter 10 Questions

1. Conduct an analysis of your personal professional development. What areas do you need to improve on?
2. Do you put your career ahead of that of your employees?
3. What formal programs should your company put in place to improve the quality of leaders and future leaders for the company?
4. What skills have you noticed that embody the concept of a serving leader?
5. Why develop employees? Who should receive training? Why? Who should not benefit from training?

Chapter 11

H5
Humor, Happiness, Health, Humility, Heart

Humor, humility, health, heart happiness, and honesty are personal traits that are not unique to supply chain leaders. Health and happiness are related, as are humor and happiness. Honesty is linked very closely to the previously discussed values/traits of integrity and ethics, and could be considered a subset of those two virtues. Humility is tied to ego. The goal of this chapter is to tie these important concepts to the discussion of modeling supply chain leadership.

Humor

I truly believe that humor and happiness are linked. *Reader's Digest* certainly believes this, as the title of their long-running series is "Humor is the Best Medicine." As a leader, it is important to model a behavior that encourages humor throughout your organization. By humor I am not talking about any form of making fun of people, or any form of humor that is offensive in nature or otherwise inappropriate in the office. I am also not talking about a form of humor that I have witnessed in several offices lately. This form of humor is where one or more employees seem to enjoy rearranging other employees' stuff while they are not in the office just to see how they respond when they return. This sort of humor was funny on the old *M*A*S*H* television series but has no place in the office.

I am certainly not speaking of the jokes and "funny" videos that seem to make their way into every office place via e-mail and the Internet. While some of the jokes and some of the videos are funny, their place is not in the office. How much time and productivity is lost every day across the country and around the world from the reading and passing of jokes via e-mail and watching videos on the Web?

Humor has a place in the office and workplace. The ability to laugh at yourself is important. Never be so serious that you cannot laugh at yourself when you do something ridiculous or stupid. One of my favorite questions at seminars and presentations on leadership is, "How many people have never done anything stupid at work?" Amazingly, no one has ever raised his/her hand. Why? Because we have all done something stupid at work.

When you do something stupid or ridiculous, what usually happens? People want to laugh. And there is nothing wrong with that — as long as the act did not hurt someone or endanger someone. When they laugh, would you rather that they laugh at you or with you? Personally, I prefer to have them laugh with me. As a leader, it is okay to laugh.

Here is an example from my experiences. In 1993, I was tasked to lead a planning team for a major exercise for the U.S. Army's V Corps in Germany. As is usually the case when an ad hoc team is put together, what I got on my team for the most part were young officers with whom other units were having problems. One particular lieutenant came with a warning from his boss that he (the lieutenant) was trouble and not a team player.

During the course of the exercise, we were required to put on our chemical suits. Actually, this was a good idea because it was wintertime and the chemical suits are very warm but not that bulky or uncomfortable. Chemical suits do not come with the names sewed on them that the regular uniforms had, so it is common practice to put a piece of green duct tape with your name and rank on the chemical suit.

As a way of breaking the ice with my ad hoc team, I put my rank, Major, and my name as "Bubba." My philosophy was that a little humor was a good idea and if it helped people to relax a little and laugh, then it was a good thing.

What was not a good thing was that overnight it rained, froze, and then snowed on top of the ice. This is never a good thing, especially for a Southern boy fresh off a series of warm-weather assignments. Walking out of the dining facility the next morning, I hit the ice and slid about twenty feet — miraculously I somehow did not spill my coffee. Out of the second-story window I heard, "Nice slide Major Bubba!" I looked up to see my "problem lieutenant" leaning out the window. When we finished laughing, he asked if I was okay and told me that he only gave me a "9.3" for the slide and that only because I did not spill the coffee.

The fact that we were both able to laugh broke the ice (no pun intended) and enabled me to establish a very good working relationship with this "problem" officer. Amazingly, I found no problems with his work ethic or his performance during the entire field exercise. It may have to do with the fact that we were both able to laugh and that I was able to laugh at myself that enabled us to establish a rapport.

Not being one to stop at one stupid thing, in a later assignment where I had more than 2000 employees, I decided to jump up in a chair to avoid the deep stuff being talked by one of my employees as a way of showing him that he was talking trash — in a light-hearted way. What I did not realize was that the chair was not stable and one of the wheels was broken. Try jumping into an unstable chair and the results will not be good. The chair flipped, I almost landed on my rear end, and my employees standing around almost lost it laughing so hard. I had a choice — laugh with them or be mad. I chose to laugh with them and years later they still remind me of the incident.

It is okay to have fun and laugh at work. As my friends in Hawaii say, "It won't break your face to laugh." Enjoy yourself and establish an atmosphere where your employees are not afraid to enjoy themselves and laugh. You may find out as I have that having fun at work leads to increased productivity. There is a direct link between enjoying what you are doing and happiness.

Happiness

Being happy with what you are doing is linked to your passion for what you are doing. We discuss passion in greater detail in Chapter 13 — but there is a link. If you are not enjoying what you are doing, how can you expect your employees to enjoy what they are doing? If you are not modeling a behavior that reflects true happiness on the job, your employees will emulate your less-than-happy attitude.

The Declaration of Independence of the United States of America states that we all have an unalienable right to "life, liberty, and the pursuit of happiness." Obviously there is a link between satisfied with the work environment and happiness. Some people appear to not have the capability to be happy, no matter where they are and what they are doing. We have all seen this type of person and this type of employee. But for the most part, if an employee works in an environment that encourages humor, encourages laughter, and where the leaders go out of their way to make the employees feel appreciated, the employees will be happy at work and will be more productive. Establishing this type of work atmosphere is a good way to prevent motivational dysfunction among your employees.

I recently worked with a company in which the only time the boss was seen was when there was a problem. The employees started to cringe like a dog that has been beaten regularly every time they saw the boss coming because they knew it was not a friendly visit. On the other hand, I watched Mike Hahn of Sprint walking around and talking with his employees on a regular basis, and he created an atmosphere where it appeared that everyone in the department was happy with what they were doing and where they were working. This was an exception during this period of time at Sprint as the company was going through some rough adjustments to the economy and was downsizing its workforce.

Your happiness is directly linked to your health, and the health of your supply chain is directly linked to the happiness of your employees and supply chain team members.

Health

In this section we look at two different aspects of health. First, your personal health and the impact it has on your outlook, and on your happiness and productivity. The second aspect is the link between employee happiness and health, and the health of your supply chain.

Why is your personal health as a leader important? What does personal health have to do with leadership and modeling leadership? There is a correlation between your health and the hours you spend in the office. Have you ever seen a person who, after having retired, wished that he/she had spent more time at work?

The balance between your time at work and your time at home or relaxing is critical to your overall health. The best example that I can provide came a few years ago from a retired general officer. During a question-and-answer period with the Advance Operational Art Studies Fellowship for the U.S. Army War College, this general gave the following guidance:

> If you have any health problems, take care of them immediately. Do not wait until you have time to take care of them. I did that and now my problems are much more serious than if I had taken the time earlier to get the problems analyzed and taken care of.

I tried to pass on this philosophy to my employees after that. One employee did not listen very well and several years later experienced heart problems that necessitated surgery and a pacemaker at a very young age. It also caused her early departure from an assignment to have the problem corrected. This same officer had extremely bad bunions on both feet but continued to run against the advice of her doctors, thus making the problems worse, and never took the time to get the problems fixed.

You cannot be an effective leader if you let personal health problems continue without action. This leads to extended absences from the workplace and does not model the behavior that is beneficial to you or your employees. If your employees emulate that same behavior, you may find yourself and your key employees incapacitated for extended periods of time. This does not contribute to the health of the supply chain organization.

More and more companies are starting to recognize the link between employee health and productivity, and corporate health and productivity. As a leader, it is important to model the right behavior and encourage employee health programs. In one assignment, I allowed my employees to spend an extra thirty minutes a day on some form of physical activity as a way of encouraging better health habits. At the same time, I tried to model the correct behavior by spending every day working out in the gym at lunchtime.

Humility

We discussed in the definition of leadership that ego is not part of the leadership equation. Leaders must be humble. It is okay, as we discuss in Chapter 13, to have professional pride in what you are doing but leaders must be careful about how much their personal pride and ego get in the way of doing what is ethical, honest, and right. Being a leader is not about personal gain — although one would wonder watching the U.S. Presidential election in 2008 or the reports of the CEOs of the failed mortgage, insurance, and brokerage firms. When a CEO takes a multimillion dollar buyout — as Bob Nardelli did with The Home Depot after tanking the stock or when a CEO takes a large multimillion dollar bonus as some of the failed mortgage companies and banks did during the collapse of the financial system in the United States in 2007 and 2008 — one has to wonder if egos got in the way.

Is there a difference between modesty and humility? Some dictionaries and thesauruses list them as synonyms. Modesty is close to humility yet you can be proud of your employees, your supply chain, and your products and still remain humble as a leader.

We discussed my colleague who told his employees, "I don't care how you do it as long as it makes me look good." Taking care of self at the expense of your employees or shareholders is a classic example of letting ego get in the way. Lead with your heart, not with your ego.

Heart

In the movie *The Replacements* after a few of the regular players had crossed the picket line to play football again, the team coached by Gene Hackman's character was asked what it would take to win the critical game. His response was, "Heart."

An online dictionary has a definition for heart that fits nicely into this discussion. Dictionary.com defines heart as "spirit, courage, or enthusiasm."[1] A leader with spirit and enthusiasm is what Gene Hackman's character was referring to in the above example. A leader who has enthusiasm, spirit of convictions, and the courage to do what is right every single time is important to every operation. A leader who models this spirit, courage, and enthusiasm will find that he/she has employees who will emulate those qualities. Employees with spirit, courage, and enthusiasm will be more productive, will be concerned about taking care of the customer, and will more than likely stay with the company even in bad times.

In the movie *We Are Marshall*, the concept of heart came into play again just like in *The Replacements*. In this case it was "heart" that allowed the entire community and the young football team to rise from the ashes of a tragic plane crash and develop the desire and enthusiasm to play the game of football again at the collegiate level. Are you modeling the right heart or desire and attitude for your supply chain team to emulate and to drive your team to new levels of success?

Summary

A leader who demonstrates his/her concern for the health of his/her employees and the health of the supply chain while being more concerned about his/her employees' success than his/her own success will be a leader for whom people will want to work. Employee retention is a good metric for benchmarking and measuring leadership.

Establish an environment where employees are happy and encouraged to have fun and laugh at work, and you will find your productivity improving and employee retention increasing.

Notes

1. Heart. Dictionary.com. *Dictionary.com Unabridged (v 1.1)*. Random House, Inc., New York. http://dictionary.reference.com/browse/heart. Accessed October 13, 2008.

Chapter 11 Questions

1. Are my employees having fun at work, or do they come to work because they have no choice and need the job?
2. What programs are available for employee health improvement, and how much does that add to satisfied employees and reduced health care costs? Am I taking advantage of these programs as a behavior model for the employees to see?
3. Am I letting my ego get in the way of taking care of the employees and improving the operations and work conditions?

Chapter 12

13
Integrity, Inspiration — Supply Chain Intelligence and Supply Chain Leadership Integrity

> I don't care how you do it as long as I look good.
> —A senior supply chain manager to his employees

Integrity, inspiration, and supply chain intelligence—what do they have to do with supply chain leadership? A leader without integrity will not be able to lead a bunch of hungry refugees to a restaurant, much less a supply chain organization. Supply chain integrity is critical to retaining customers and profits. This chapter looks at areas in the supply chain that impact an organization's supply chain integrity. A leader without integrity may also find that he or she cannot inspire their employees to accomplish the goals of an organization and is of little value to the employees or the company. Part of the leadership definition from Chapter 1 includes providing motivation to employees, a leader that cannot provide inspiration to excel will not be able to provide motivation—he or she may fuel motivation dysfunction, but will not motivate his or her employees.

This chapter introduces a new concept that impacts the integrity of supply chains: supply chain intelligence. In military operations, intelligence is important to the planning of future operations. Intelligence in the military includes knowing what the enemy is doing and what the allies are doing or are capable of doing. Supply chain intelligence is knowing what is going on in the supply chain to include what the competition is doing or is capable of doing to provide better support to the customer and what our supply chain partners are capable of doing to enhance our support of the customer. In supply chain operations, supply chain intelligence is critical to success and helps set the conditions for success.

Integrity

What is *supply chain integrity*? Is there a link between corporate success and supply chain integrity? This chapter looks at the concept of integrity from a leadership perspective and the link between leader integrity, supply chain integrity, and the success of an organization.

The handling of the books prior to release of the last two Harry Potter books demonstrates both good and bad supply chain integrity. The sixth installment of the Harry Potter books was carefully controlled and delivered in time for the midnight start of sales. The final installment of this successful series somehow was leaked to the press before the official release date and reviews of the book appeared before the book was supposedly available for release. How much integrity is there in your supply chain?

Like ethics, it is important to define integrity in order to establish a baseline to measure, model, and benchmark. Integrity forms one of the cornerstones of leadership and must be demonstrated every day. Integrity is one of those virtues that many dictionaries have trouble defining. It is one of those virtues that is easy to recognize and sometimes hard to completely define. Integrity is acting on a set, a consistent set of values that is consistent with established cultural principles or mores.

A U.S. Air Force manual on leadership defines integrity as the "ability to hold together and properly regulate all the elements of one's personality. A person of integrity acts on conviction, demonstrating impeccable self-control without acting rashly."[1] The manual goes on to state that "Integrity is not a suit that can be taken off at night or on the weekend or worn only when it is important to look good. Instead, it is the time that we least expect to be tested when possessing integrity is critical. People are watching us, not to see us fail, but to see us live up to their expectations of us." The same is true for your organization, regardless of what your organization does. People are watching you to see how you act and to see if you are living up to their expectations of your integrity.

Benchmarking integrity is easy; like ethics, it is a go/no-go measurement. Either you, your company, or your employees act in a manner that is consistent with established principles — or you don't. There is no in-between. Modeling integrity is also easy. Setting the example for integrity is a daily responsibility for leaders at all levels of the supply chain.

Developing employee and customer loyalty and retention is linked to the integrity of the supply chain leadership of a company. Even the integrity of your spokesperson may impact your customer loyalty and retention. Look at how quickly a company drops a spokesperson after an incident that leaves the company with questionable integrity. Prominent sports figures come to mind when looking at this principle. Michael Jordan continues to be a spokesperson long after his playing career finished because he is perceived as a person of integrity. Likewise, Michael Vick, Marion Jones, and O.J. Simpson were quickly dropped as spokespersons when their credibility came into question.

Does your supply chain integrity make your company the weakest link in the supply chain? Unless you are the sole source — not simply the single supplier, actually the sole source for the item — a lack of supply chain integrity and supply chain ethics could leave you as the weakest link in the supply chain. If you are deemed the weakest link and you are not the sole source of the item, you may very well find yourself in the situation of many of the contestants on the popular British game show *The Weakest Link*. The last thing you want to hear from your customers is the line from this show: "You are the weakest link. Good-bye!"

Part of the modeling of supply chain integrity is the need to sample products and measure your supply chain partners. Supposedly, Mattel depended on long-time suppliers to test its products and accepted the results. Another major business supplier depends on its supplier to ship "quality" products from China — but has discovered that some of its trusted suppliers were actually shipping defective products, resulting in this major supplier having to inspect samples of every lot coming from that supplier. This action was taken to protect the integrity of the major supplier.

Supply chain integrity is on display every day in every organization. In 2007, the integrity of several supply chains came into question. How a company responds to integrity issues speaks volumes about that company.

Integrity is a measure of the believability of your actions and your words. It is a measure of "can I trust this person or this company?". Integrity is one of the foundations of business and leadership. Without corporate integrity, customers will not want to do business with you. Without personal integrity, employees will not want to work for you, will not trust you to do what you say you will do, and will most likely start looking for a new place to work — or worse, will start imitating your actions and the integrity of your company will become suspect from the perspective of your customers. When customers begin doubting your integrity, especially in the supply chain industry, they will start looking for another supplier and eventually your company will lose business. Integrity is another one of those all-or-nothing metrics of your character and the character of your company. Integrity breeds trust and confidence in you and your supply chain. Regardless of your position in the supply chain, customers want to do business with a company they can trust, and trust comes from unquestionable integrity.

Supply Chain Integrity and Personal Integrity

Personal integrity is the first step in developing supply chain integrity. Although the Osmond Brothers had a big hit in the 1970s with "One Bad Apple (Don't Spoil the Whole Bunch Girl)," in business one person with a lack of integrity can indeed spoil the whole company. A leader with an integrity problem creates one of two situations. The first is that customers and employees recognize the lack of integrity and vote with their feet to move on. The second is perhaps worse — employees start to emulate the actions of the integrity-challenged leader and start to act in a manner that exhibits a lack of integrity, which in turn drives away more customers.

Integrity is easy to model for employees. Always conduct your actions in a manner that will not even give the illusion of a lack of integrity; and not only will you be able to look yourself eye to eye in the mirror, but your employees will emulate your actions and conduct business in a manner that exudes integrity.

I once worked with a very talented officer who unfortunately did not have integrity, honesty, or any professional ethics. This officer would make excuses as to why something could not be done or would constantly make excuses to his wife as to why he could not come home on time (to avoid taking care of his children). The classic demonstration of his lack of integrity and honesty was displayed as he prepared to leave for a new assignment. His father had written a very successful book on logistics and the officer's fellow employees would bring copies of his father's book to be signed when his father visited. I noticed a stack of these books on the officer's desk and asked him about them because he would be leaving soon. His response floored me. He said, "Don't worry. I can sign his name as good as he can; so if he doesn't make it back, I'll sign them for him."

Another officer who worked for me was the son of one of my former bosses. I once commented that his signature looked just like his dad's, which was strange because they had different first and middle names. His response also floored me and left me questioning everything that he said after that. His response was, "I practiced signing his name and signed so many documents for him while I was in high school that I guess I mastered it and now mine looks like his." How can you trust someone like that to lead employees? I am proud to say that he was asked to leave the U.S. Army because of his lack of integrity.

Personal integrity is important for leaders because you are being watched by all of your employees. You have to conduct business with that in mind because if your integrity is questionable, your ability to deal with personnel will be impaired. The higher you climb the leadership ladder, the more the problems that you face will be interpersonal. By the time you move into a leadership position, in most cases, your technical skills are no longer in question. However, behavioral issues take the forefront. It is in these behavioral and interpersonal problems that you must make sure that your integrity is not in question. Once you compromise your integrity, your ability to be a leader is compromised. And unfortunately, it only takes one act to wipe out a career of integrity.

This concept surfaced recently concerning the purchasing activities in Kuwait during the build-up and early days of Operation Iraqi Freedom. News reports during the latter half of 2007 detailed the activities of a few soldiers who took bribes and kickbacks to award contracts. One officer involved his wife and reportedly pocketed close to $1 million before the actions were discovered. I am sure that, based on the ranks of the soldiers involved, they had more than likely demonstrated integrity during their careers to get to the positions they held at the time of the alleged incidents.

The same is true of Enron. The senior executives of Enron all had successful careers in business before they joined Enron, yet somehow their integrity compass became demagnetized by the lure of the riches of the energy business.

Everyone is born with a compass that points true North for ethics and integrity — somehow some people get their compasses demagnetized and the result is actions that leave others questioning their integrity. The poem "The Man in the Glass" by Dale Winbrow has been one of my favorites since I first read a laminated copy of it that I found in my Dad's desk drawer as a teenager. I have kept a copy of the poem on my desk for more than twenty years as a guide to integrity. One line in the poem states, "The man in the glass says that you're only a bum, If you can't look him straight in the eye." This is the true measure of integrity as a supply chain leader — always act in a manner that will allow you to look the man/woman in the glass straight in the eye.

Never make excuses — never lose faith in yourself or your employees.

Can you have a lack of personal integrity in your supply chain without having a supply chain integrity problem? It is possible — but not likely. Breaches to supply chain integrity can result in irreparable harm to the confidence placed in your supply chain. The impacts could be revenue loss and loss of customer confidence, both of which could drive customers to other suppliers. Supply chain leaders are responsible for ensuring that their risk assessment and risk management strategies are linked to the total supply chain, from the planning stage through to the delivery stages of the supply chain.

Supply chain integrity involves getting the right product to the right place, at the right time, in the right condition, and now getting a product to the customer that is safe, even if not used in the manner for which the product is intended.

Supply Chain Integrity and Forecasting

Can your forecasting impact your supply chain integrity? If supply chain integrity entails getting the right product to the right place in the right quantity, then forecasting errors can definitely impact supply chain integrity.

Here are some examples of improper forecasting impacting supply chain integrity. At the 2008 Daytona 500, billed as "the most anticipated event in motor racing," the programs were sold out hours before the race started; before the race was half over, the concession stands were sold out of Coke® and peanuts.

2007 was a record year for the University of Kansas Jayhawks football team. However, the concessions company had forecast consumption rates using the 2006 data. Under normal circumstances, this would have been an acceptable practice — but this was not a normal circumstance. When a team is suddenly winning all its games by huge margins, more fans start coming to the games. Two football games that did not sell out in 2006 were sold out in 2007. The impact on the supply chain? In the third quarter all of the concession stands sold out of pretzels, hamburgers, and hot dogs. The fans did not want to hear that the forecast was based on the previous year's data; they just wanted food and drinks.

There is nothing more embarrassing to a company than to under-forecast the appeal of its product, as IBM did several years ago with its new laptops, or over-forecasting and resulting in excess product that must be sold at discount prices.

Supply Chain Ethics and Integrity

According to a White Paper by TransportGistics, "Supply Chain Integrity describes a chain that is sound and free of corrupting influences. Its size and scope ranges across the entire commercial enterprise."[2] The longer supply chains become, the harder it becomes for supply chain leaders to ensure the integrity of their supply chains. This is important from both a risk management perspective and a risk assessment perspective. Supply chain leaders are responsible for conducting a thorough risk assessment of their operations. Part of this risk assessment must look at the vulnerabilities of the security of the supply chain.[3] Another aspect of risk assessment is to look at all the links in the supply chain for leaks of products into the gray market, thus lowering the value of the products in the legitimate and intended markets. These gray markets may include permanent flea markets and Internet sites, as well as outlet malls and kiosks.

In 1984, there was scare with Tylenol in the Midwest because of the discovery of bottles that were tampered with and supposedly laced with cyanide. McNeil Labs/Johnson & Johnson quickly reacted to the situation and removed all the products from the shelves and quickly replaced the "tainted" products with new tamper-resistant packaging that has become the industry standard for pharmaceutical products. Because of quick action, the company was able to fix the problem while keeping consumer confidence in the product and improving the safety of all over-the-counter pharmaceutical products. One such bottle of Tylenol made it to my predecessor in Company Command at Fort Gordon, Georgia. One of the last things that my predecessor told me was that there was a bottle of the tainted lot of Tylenol in the top right-hand drawer of my desk and if things got too bad, to take a couple of them. I passed the bottle on to my replacement two years later with the same words of advice.

Is supply chain integrity really a problem in today's business world? Just as there are numerous examples of a personal lack of integrity, there are also numerous examples in the past year or two that reflect a supply chain integrity problem.

Another example of impacts of supply chain integrity surfaced at Daytona in February 2008. According to the NASCAR.com Web site, Jimmie Johnson, Casey Mears, Dale Earnhardt, Jr., Jeff Gordon, Denny Hamlin, A.J. Allmendinger, J.J. Yeley, Scott Riggs, Clint Bowyer, and Tony Stewart were forced to change their engines prior to the start of the Speed Weeks Dual 150 Mile races due to a faulty coating on a lifter (a cylindrical part that rests against the camshaft lobes and causes the valves to open and close while the camshaft rotates). The chief engine builder for one of the teams affected by the bad coating stated that he believed the lifters "came from a bad batch received from the company's supplier." According to another engine builder, the defective coating could cause the steel lifters to become like glass and shatter during the driving of the race cars.[4]

> If you do things the way you've always done them, you'll get the same things you've always got.
> —Darrell Waltrip

There is a standard answer in distribution centers around the world when asked, "Why are you doing things that way?" The answer is almost always, "We've always done it that way." Just because it has always been done a certain way does not make it the right way. Doing things the wrong way will impact the integrity of your supply chain. The old Oldsmobile commercial used to claim that "This is not your dad's Oldsmobile." Obviously this commercial was a while ago as Oldsmobiles are no longer made. The supply chain of today is no longer "your dad's supply chain." When everything was sourced locally and produced locally, supply chain integrity was not a key issue. With the lengthening of supply chains and outsourcing becoming more prevalent, the words of Darrell Waltrip warn us that to maintain supply chain integrity, all of us in the supply chain business must change the way we think about supply chain integrity and adopt a new mind-set for doing business in a global environment. I recently saw a quote, attributed to Alan K. Simpson, that stated, "If you have integrity, nothing else matters. If you don't have integrity, nothing else matters." The same is true for supply chain integrity.

Supply chain integrity problems surfaced in the summer of 2007 when Mattel had to recall toys containing lead paint. And then again the supply chain integrity issue surfaced in the fall of 2007 in the toy industry when a recall was necessary for toys with dots that replicated the date-rape drug GHB when accidentally swallowed by small children, putting some of them into coma-like conditions.

Part of your supply chain integrity depends on sample testing your products, regardless of where they are sourced. Could the recalls of Mattel and others have been avoided with product testing and sampling? More than likely the answer is yes. One major wholesaler tests products from China after they are delivered to its Kansas City distribution center. This particular wholesaler has discovered from its sample testing that some of the products from certain suppliers had to be 100 percent tested before shipping to retail locations. Is it better to test a product that flies in the face of certified suppliers and quality initiatives than to have a national recall? Does the lack of supply chain integrity impact your corporate reputation? Absolutely!

Mattel allowed long-time suppliers to test themselves. Apparently this practice does not always work. Another FORTUNE 500© company allowed its third-party transportation provider to measure itself. Amazingly, the 3PL reported 99.5 percent on-time delivery every single month and no one seemed concerned that in heavy months and in light months the percentage of on-time deliveries never deviated. Can you allow your 3PLs to sample, test, and measure, and then grade themselves on critical metrics and customer service? Admittedly, there is a requirement for trust in a supply chain between partners in that supply chain. However, the bottom line is that when a defective or hazardous product is delivered to the customer, it is your company's name and supply chain integrity that is on the line and potentially damaged.

Sometimes the lack of supply chain integrity impacts the delivery or completion of a product. Boeing learned this the hard way. In 2007, Boeing rolled out its new 787 super airliner. The much-ballyhooed test flight was delayed and the first model was simply rolled out for the announcement. Why? The test flight was delayed due to the inability of supply chains (a lack of supply chain integrity) to provide some of the key fasteners — the rollout model used temporary fasteners that were not flight-worthy.

According to *Logistics Today Magazine* (February 2008 edition), Boeing again delayed the delivery of its fuel-efficient 787 plane. This time the delay was for more than a year. Boeing cited supply chain problems as the cause of the delay. Probably more than just a coincidence in this case is that this was the first time that Boeing "had major components built in other countries and then sent to Everett, WA for final assembly." The integrity of the supply chain has a trickle-down effect. In this case, the trickle-down effect impacted All Nippon Airways and Qantas Airways, which were planning on using these planes in 2008.

Does your supply chain integrity make you the weakest link in your supply chain? Can you be dismissed as easily as contestants on the popular TV show *The Weakest Link*? Will your lack of supply chain integrity force your customers and partners to say, "You are the weakest link. Good-bye!" as the host of the TV series did to contestants voted off the show?

As of 2004, there is an International Standard Organization standard to ensure the integrity of the food supply chain.[5] This ISO 22000 Standard came about because "failures in food supply can be dangerous and cost plenty." The standard is "intended to provide security by ensuring that there are no weak links in the food supply chain." The intent is to ensure the integrity and security of the food supply chain to improve customer satisfaction and customer confidence in a vulnerable supply chain. This standard was developed and approved based on the premise that supply chain integrity promotes supply chain confidence. Supply chain consumer confidence can translate into lower inventory levels, which produces reduced investment in inventories while improving customer service. Improved customer service leads to happier customers, which usually leads to greater profits and additional reduced costs to attract new customers.[6]

Just as the food supply chain is concerned about supply chain integrity and customer confidence, pharmaceutical supply chain leaders are also concerned about the integrity of their supply chains. Pharmaceutical supply chain leaders face the constant concern of delivering safe and effective medicines from raw materials to finished products for patients. A break in the pharmaceutical supply chain could force supply chain leaders to seek alternative sources that require certification, testing, and sampling to ensure product purity.

Integrity in Purchasing and Acquisition

A major Department of Defense contractor that I interviewed with prior to retiring from the Army bragged about a party it threw, at government expense, for all its employees. The storyteller spoke of how the party ran up a bar tab of more than $10,000 at this particular "meeting." My instincts told me that although this company supposedly had a good reputation in the contracting world, that this lack of integrity meant that this company was not a good fit for me in a new career.

Dr. Rob Handfield of the North Carolina State University Supply Chain Resource Consortium wrote an article a few years ago on the link between ethics and supply chain integrity in a global environment. The article included this guidance: "Your reputation for integrity will serve you well."[7] This article got me thinking about the link between supply chain integrity and purchasing.

A major Department of Defense nonprofit support organization holds an annual convention every fall. The convention is known for its large corporate-sponsored booths and parties for senior officers. These booths and parties are replete with gifts, food, and liquor. The sole purpose of this convention is to influence the buying habits of senior military officers for contracting services and major equipment, and to thank the officials for their business. The amazing part of this example is that the same officers who attend this convention are quick to question the integrity and ethics of attending supply chain trade shows. One went so far as to comment, "The vendors at the trade show that you are scheduled to speak at are out to influence you to buy their products." My response was, "If that is unethical or an integrity violation, how do you perceive the XXXX Annual Convention since that is the sole purpose for that convention?" I was cleared to speak at the trade show.

What about different standards of ethics and integrity in different societies and countries? Does maintaining your personal integrity impact your company's supply chain integrity? Different cultures and different societies have different standards of integrity — before you start jumping on a soapbox and preaching the understanding of cultures, bear in mind that my experience includes working in some countries that most of us would never do business with or even want to visit. My statement about different standards of integrity is not a judgment of other countries — it is a simple statement of fact. The payment of what some would call bribes in the United States is considered a common practice in some societies — even in the United States. The acceptance of gifts is a standard practice in many cultures. Does this make it right or wrong? Absolutely not. What makes it wrong is if the gift is of such value that the acceptance of it gives the illusion of impropriety or swaying your decision to go with that particular vendor or supplier.

The U.S. Government has very strict rules about accepting gifts, meals, or drinks from vendors. Why? Because the Government does not want to have its contracting personnel providing the illusion or impression that there is any favoritism based on gifts. Because of the number of military personnel who transition to contracting jobs after retirement, this sometimes creates a problem when old friends get together, but the key is to prevent the impression that any favoritism is granted because of gifts, meals, or drinks.

The moral here is that there is no such thing as situation ethics or situational supply chain integrity. Integrity is an all-or-nothing measure of leadership, regardless of the business that you are in. Your employees are watching you to model integrity for them to emulate.

Can you maintain your supply chain integrity while continuing to outsource non-core competencies and manufacturing? Look at the example of Disney. The Walt Disney Company recently discovered that a company that had not passed Disney's audits and was supposedly removed from Disney's supply chain was, in fact, still being used by another contractor in their supply chain in China and was not following the prescribed labor standards, thus resulting in a lawsuit.

Maintaining supply chain integrity becomes more important as supply chains become longer and more global. As the supply chain becomes more global and more extended, the control over the supply chain becomes less.

Part of the supply chain integrity equation that is growing in importance concerns the working conditions of the companies that form the links in the supply chain. One of the most effective methods of ensuring adequate working conditions and work hours for supply chain partners is to conduct routine audits of the operations of your suppliers. However, as we saw in the Disney supplier problem, conducting an audit may not be enough. What is necessary is unannounced audits of supplier working conditions and audits that focus on continuous supply chain improvements.

There is a direct link between cargo and port security, and the integrity of your supply chain. Supply chains are inherently complex, dynamic, and fluid. Global supply chains are characterized by uncertainty, ambiguity, and potential interruptions in the supply chain. These characteristics cloud the operating environment and they create risks.

These risks were highlighted after the terrorist attacks of September 11, 2001, when estimates of losses to the economy reached $2 billion a day because nothing was moving throughout the United States. Less than a year later, the longshoreman's strike on the West Coast once again stalled the supply chains as 300 to 500 ships sat idle off the coast of California.

According to George C. Mathy in a *Logistics Today Magazine* article (January 16, 2008), "7 Tech Trends for Logistics," "U.S. Homeland Security officials have warned that the next terrorist attack could be launched through the country's cargo and port systems." When you consider that 58 percent of the containers coming to the United States come through three major ports and that approximately 12 million containers come into the United States alone, or that approximately 200 million containers are moving around the globe each year, the integrity of supply chains obviously becomes tied to port and cargo security.

Another aspect of supply chain security that impacts supply chain integrity is the shrinkage of items from distribution centers' back doors. The focus since September 11, 2001, has been on external impacts on the supply chain.[8] This has opened the door, literally, to employee theft rings. Estimates for losses from employee theft range as high as $60 billion annually in the United States alone. This ties employee background checks, internal security, and employee integrity to supply chain integrity.

Successful risk management promotes supply chain integrity. According to the Aberdeen Group in a paper on supply chain risk, the "Best in Class enterprises are 54 percent more likely than their peers to identify supply risk as a high priority for corporate action."[9] Corporate risk assessment and risk management programs are necessary for companies to maintain supply chain integrity.

When doing risk assessments and developing risk management programs, it is important to look at some of the following areas:

■ *Vendors and distributor networks.* Are your vendors and suppliers conducting a risk assessment and auditing their suppliers?
■ *Supplier certifications.* Are you using certified suppliers? Are you spot-checking your certified suppliers?
■ *Employee background checks.* Are you and your suppliers doing valid background checks on employees? Is there a problem with defective parts or shrinkage in your distribution center because of a lack of employee background checks?
■ *Routine cycle counts of high-value items.* After completing a risk assessment of your operations, do you find that there is not enough security for your high-dollar items? Would a more frequent cycle count help control the shrinkage, or is a more secure area in the distribution center required?
■ *Routine sampling of raw materials and purchased components.* Are you conducting a routine sampling of your raw materials, purchased components, and finished goods before shipping them to your customers?

There is a direct correlation between supply chain security and supply chain integrity. Would you rather do business with a safe supply chain with integrity or an untested link in the supply chain simply because of lower cost? Your supply chain integrity will determine if you are the weakest link or the strongest link in the supply chain. The choice is yours and yours alone. A supply chain leader provides vision, direction, motivation, and sets the example of integrity for employees to emulate and follow.

How you handle customer mistakes also reflects on your supply chain integrity. I mistakenly ordered the wrong color of gloves for my daughter for homecoming. I contacted the company and explained the problem. That day, a new pair of gloves was expedited in time for that weekend's dance with the agreement that I would return the other gloves in the mail. The new gloves were actually more expensive but I was not charged any additional costs. Guess where this year's gloves will be ordered!

Supply Chain Integrity and Reverse Logistics

Is there a link between supply chain integrity and reverse logistics? Isn't that stretching it a bit too far? Not really. There is a direct correlation between returns and supply chain integrity — and not just the returns dictated by recalls as discussed earlier. There is also a perception that the integrity of your supply chain and the way you handle returns are related.

What do your returns tell you about the product — quality, delivery, on time? Reverse logistics, according to *Forbes Magazine* (November 3, 2005), is a $100 billion industry in the United States alone. U.S. companies have created a try-it-before-you-buy-it mentality with their liberal return policies. However, there is still a large industry in processing returns that come back because of faulty products. How you handle returns and get a quality product in the hands of your customer reflects on your supply chain integrity.

Some products come back because of customers ordering "excess" products due to delivery integrity violations. This happens when a promised delivery is not met or there is no faith in the supply chain.

This happens on a regular basis in the military. When a repair part is ordered, the information systems are updated on the status of the order. The common practice for repair parts clerks when asked about an order is to check the system and reply, "I've got good status." As soon as the young maintenance officer walks away content that the order is good, the repair parts clerk orders another one. The result is several parts on order for the same repair job. When they all come in, the excess parts are sent back through the reverse supply chain. L.L. Bean experiences the same phenomenon with shirts. They realize that if a customer orders three shirts, one of them will probably come back because the color does not look good on the customer.

Reverse logistics problems are not new in the United States. The end of the American Civil War provides an example of a way of not handling problem inventory. After the surrender of the forces of the Confederate States of America, General William T. Sherman was faced with moving north across a swollen Neuse River outside of Raleigh. According to North Carolina lore, unwilling to carry unneeded materials and ammunition forward, General Sherman, or at least General Sherman's soldiers, took the action that became the standard for returned merchandise for almost a century — dump it. Since landfills had not been invented yet, General Sherman's soldiers simply dumped the unneeded stuff in the river and on the banks of the river to make the crossing easier.

Montgomery Ward started its policy of "If you are not satisfied, bring it back for a full refund" in 1894. This policy continues today in many companies, prompting a try-it-before-you-buy-it philosophy in today's consumers.

Why does a product come back through the supply chain? What is driving your returns? Discounting the obviously negligently damaged products and the abuses of the returns policies, what is coming back can give supply chain leaders a good idea of problems in the manufacturing and supply chain processes. Analyzing the returned merchandise can give supply chain leaders a good idea of where there may be a breakdown in the integrity of the supply chain.

One executive in a major company told me that the company does not worry about returns because there is no money in returns and the return rate is so low. Having been in a number of this company's stores, I am not sure that this high-ranking executive was in touch with reality in the business. One major retailer in the United States averages about $6 billion annually in returns, which equates to approximately forty-six trailers a day, every day going backward with returned merchandise. Every company has a problem or opportunity because of returns. The Reverse Logistics Association defines reverse logistics on their Web site (www.reverselogisticstrends.com) as "all activity associated with a product/service after the point of sale." The Supply Chain Council (www.supply-chain.org) defines returns as, "processes associated with receiving returned products for any reason. These processes extend into post-delivery customer support." The link between these definitions and supply chain integrity is the reason for the returns and how the company handles the communications with the customer.

Look at the statistics for returns in the United States in Figure 12.1. The first question is, "What is coming backward?" Every company in the supply chain business needs to evaluate what products are coming back from the customer. The second question in the integrity process is, "Why are these products coming back to us?" The next question is, "What is the impact on our supply chain integrity because of the defective products?" The final question is, "What are the impacts to other areas of our supply chain because of these integrity violations?" See Figure 12.2.

Supply chain integrity can be impacted by problems in any one of the areas listed in Figure 12.2. When problems with items coming backward in the supply chain are not quickly handled to the satisfaction of the customer, these problems become larger.

Reverse Logistics Data

- \>$100 billion worth of goods are returned each year according to forbes magazine and the reverse logistics association
- U.S. Companies spend >$45 billion on processing returns
- Approximately 5%–30% of sales end up as returns–led by high tech sales
- 25% of christmas season sales come back to retailers as returns

Figure 12.1   Reverse logistics data.

Impacts of Reverse Logistics
- Forecasting
- Carrying costs
- Processing costs
- Warehousing
- Distribution
- Transportation
- Personnel
- Marketing

Figure 12.2   Impacts of reverse logistics on supply chain operations.

Supply Chain Integrity and Delivery Integrity

Some of the problems discussed above concerning returns may be a result of delivery integrity and the resultant customer dissatisfaction. Violations of delivery integrity also manifest themselves in missed deliveries. I used a major flower delivery company for years until promised deliveries were missed. A couple of years ago, I ordered flowers for my wife's birthday and there was no problem — the flowers were delivered as promised. Three weeks later, I again ordered flowers for our anniversary. I took off work early that Friday to be home when the flowers were promised according to the Web site information. When I arrived home, there were no flowers. I called the company and was told that the address did not exist. Imagine my surprise to find out my address did not exist. When I explained that they had delivered to the same address three weeks earlier, I was told that the truck had already headed back to the terminal and that they could deliver the flowers on Tuesday (Monday was a holiday) or they could credit my account for the flowers. My guess was that fresh flowers probably would not be fresh after sitting in a distribution center in warm weather for an extra four days. I have since used a different vendor for flower deliveries.

Supply Chain Integrity and Organizational Design

Is there really a link between how your company is organized and your supply chain integrity? How does your organizational design impact the integrity of your supply chain? Traditionally, organizational designs had supply chain functions under several different departments, causing competing demands and competing interests for personnel, inventory levels, and financial investments. A traditional organizational design looks similar to Figure 12.3. Because there are multiple departments, multiple vice presidents/directors, and competing/contrasting goals and objectives, this form of organizational design can and usually does produce fragmentation and potential impacts on the integrity of the supply chain.

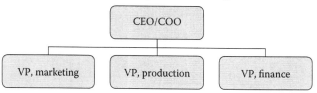

Figure 12.3   Traditional organizational design.

One FORTUNE 500© company that I recently worked with had this type of organization. As a result, the returns and reverse logistics programs were under one vice president, and the forward supply chain was under a different vice president. The result of this organization was a degradation in customer service, a disconnect between the forward and reverse supply chains, and a contract that rewarded their 3PL providers more for processing returns than for processing sales and forward supply chain shipments. Another FORTUNE 500© company fragmented its operations across several departments on the same contract, resulting in impacts on customer service and fragmented guidance to the employees. Fragmentation of functions can result in lost orders, slow processing of returns, or loss of customer confidence in your supply chain.

A more functional organizational design is emerging as more companies adopt a functional supply chain model. This organization as depicted in Figure 12.4 places

Figure 12.4   Functional organizational design.

Supply Chain Integrity
- Perfect order fulfillment
- Information security
- Parts integrity
- Product integrity
- Shipping integrity
- Inventory integrity
- Cornerstone of supply chain leadership

Figure 12.5 Supply chain integrity summary.

all supply chain functions under one leader, making one person in the organization responsible for all supply chain functions. The advantage of this kind of organization is centralization of responsibility. Placing one person in charge and establishing the proper supply chain metrics enhances customer support, streamlines operations, and supports supply chain integrity, as depicted in Figure 12.5.

Supply Chain Integrity and Supply Chain Fraud

What practices are in place to prevent fraud in your operations and supply chains? What is fraud? Fraud is a purposeful deception, misrepresentation, or concealment of facts intended to cause injury or loss to another party, typically for one's own direct or indirect gain.

According to a Katzscan report on supply chain fraud, "corporate revenue is reduced by 3 percent per year due to fraud."[10] Supply chain fraud is happening in the public and private sectors of business. Norman Katz defines supply chain fraud in his White Paper on supply chain fraud as follows: "the term 'supply chain fraud' encompasses a growing threat that strikes both wide and deep at an organization's operations along the internal and external aspects of the supply chain."[10] Supply chain fraud impacts your supply chain integrity. Purchasing fraud, intentional or unintentional, can result in substandard raw materials or components. Internal fraud may result in inferiorly manufactured products or defective products. This impacts the integrity of your product and the customer perception of that product. External fraud may impact your supply chain integrity by putting fraudulent products on the market that may not meet your company's standards for quality.

The Sarbanes–Oxley Act mandates that companies must perform a risk assessment across all operations. It should not take legislation for companies to conduct a risk assessment of their operations. All companies should conduct risk assessments of all operations to ensure supply chain integrity. The use of job books and process maps helps communicate requirements and assists in identifying areas of risk or potential fraud in your supply chain while educating employees on what right looks like so they can identify fraudulent activities when they occur.

Supply Chain Integrity and Flow/Distribution Center Layout

Does the layout of your distribution center really have an impact on your supply chain integrity? What happens when the flow through the distribution center is not properly planned, or the plan is not followed? Experience shows that when the flow through the supply chain is not carefully planned or not followed as planned, items disappear, or get misrouted or damaged. When items get misrouted and the resultant misdelivered or damaged item is delivered to the customer, confidence in the integrity of the supply chain is damaged. During the build-up stages of Operation Iraqi Freedom, there was a problem with items not getting to the intended customer. There was no fraud in the supply chain, there was no intentional redirection of supplies, and there certainly was no pilferage of supplies.[11] But the supplies sent to Kuwait to support units and soldiers had the potential of being handled more than twenty times before being delivered to the intended unit. When an item is handled that many times, there is the possibility of damage or misdirection, and delivery to the wrong customer.

However, this is not a problem unique to the military. Several years ago, Gillette invested millions of dollars in radio frequency identification (RFID) tags for their Mach 3 razors and blades. Why? Because the shipments of the razors were not arriving at the intended locations, and Gillette was concerned that the razors were being stolen or hijacked between the distribution center and the customer stores. The RFID tags showed that the reality was mis-shipments and not stolen products.

However, the result was the same. The result was a loss of confidence in the supply chain because of a supply chain integrity problem. Regardless of the internal workings of Gillette, delivering less than promised reflects on your supply chain integrity.

Supply Chain Integrity and Risk Assessment

What about risk assessment and risk management to prevent supply chain integrity problems? What is risk assessment? In the supply chain, it should include the proper supplier and vendor metrics, and the use of certified suppliers and supplier audits. Although as Disney discovered recently, even that sometimes does not solve the problem. Disney was cited for products that were produced by a supplier with which Disney thought it had severed ties because of violations of Disney's supplier codes. However, that manufacturer started doing business with another Disney supplier working out of Japan. Disney revoked the license from Hao Wei Plastic Manufactory in 2001 and although it maintains a supplier database for its 40,000 suppliers and 6000 licensees, this problem surfaced in 2007.

Responsibility and accountability are tied to supply chain integrity. The question becomes: how far are you responsible and accountable for the violations of integrity in your supply chain? If the Supply Chain Council's Supply Chain Operations Reference Model is accurate, and it certainly seems to be comprehensive and accurate, your supply chain goes from your suppliers' suppliers to your customers'

customers. This being true, then where does your responsibility for ethics and integrity in your supply chain end? If your name will appear on your product or service, then you are responsible for actions throughout the entire supply chain.

In the March 11, 2008, edition of *Global Supply Chain News*, Dan Gilmore wrote:

> *Supply Chain Digest* reported last year how Wal-Mart, Land's End, Nike, Liz Claiborne, The Gap, Target, and more, a virtual "who's who'" of retailers and apparel marketers, were caught up in a mini scandal involving major pollution by a Chinese textile manufacturer. The issue: those U.S. companies were buying t-shirts from one company that was sourcing fabric from the polluting company elsewhere in China.[12]

Hopefully, these stellar companies were not knowingly buying from a company that was known to be a major polluter. But the question remains: How far does your supply chain integrity reach? Do the actions of your upstream suppliers reflect on the integrity of your supply chain? Should this have been identified in the risk assessment?

Supply Chain Integrity and Data Integrity

On April 4, 2008, Advance Auto Parts had information lost to a hacker. This information included the employee information for 56,000 employees in fourteen states. This type of data/information loss can certainly impact the confidence of the employees that the company will protect its information. The same is true for the customers of this supply chain. A loss of supply chain information confidence is tied to your supply chain integrity and your supply chain leadership.

Will RFID tags improve data integrity and therefore supply chain integrity? Wal-Mart believes so, as does the Department of Defense. Wal-Mart mandated the use of RFID tags to add visibility to the supply chain, improve data acquisition and accuracy, and of course improve profitability and lower prices. The Department of Defense started using RFID tags in 1994 as a test to improve supply chain visibility and prevent the problems experienced during Operations Desert Shield and Desert Storm when 27,500 ISO containers were sitting on the dock with no idea of what were in them or to whom they really belonged. This is not a situation that adds confidence to your customers. The Department of Defense mandated the use of passive tags in 2005 to improve the integrity of the data and visibility of items moving through the Defense supply chain. Here is an example of the use of RFID tags in improving data integrity while improving customer support.

Lockheed Martin, in support of the U.S. Air Force, averages more than 50,000 transactions per month. Every item in this supply chain is tagged with RFID tags. Lockheed Martin has calculated that it costs in excess of $13,000 if the line goes down for a parts stockout. Based on the use of RFID tags to improve data integrity, the company has achieved a fill rate of 99.5 percent for its 90,000 SKUs in this program. The Department of Defense RFID mandate was the impetus for this program. The use of RFID tags provides this defense contractor with traceability of all parts in support of this aircraft repair/rebuild program. Why is this traceability so critical? The supply chain integrity of parts for the aircraft repair industry and the need to trace parts origin and use to prevent aircraft crashes. The integrity of the supply chain is improved by improving supply chain data integrity.

Supply Chain Integrity Summary

Most companies revert to setting the tone for integrity and fraud prevention by publishing policy letters and corporate directives. There is more to setting the tone than just writing policies or directives. Supply chain integrity must be modeled by leaders for employees to emulate.

In war, one of the biggest concerns of leaders is preventing violations to the Law of Land Warfare. These violations can wipe out all the goodwill created by the U.S. military in the area in which it is operating, as well as wipe out goodwill feelings in the international community. Just as there is a requirement for the U.S. military to investigate violations to the Law of Land Warfare, you must also investigate violations of the integrity of your supply chain. How do you handle integrity violations in your supply chain? How you handle these violations, much like McNeil Labs/Johnson & Johnson did with Tylenol, will impact how the public views your company and whether or not customers return to your company or "vote with their feet." How you handle mistakes in your supply chain reveals the strength of your supply chain integrity and resolve to maintain your integrity and not try to practice situational integrity.

The test for integrity and ethics is very simple. Can you look yourself in the mirror the morning after the big deal is closed or the day after the product ships? Would you knowingly sell a defective item or defective part? Some companies have shown that they could both sell a knowingly defective product and still be able to look themselves in the mirror. A person of integrity would not knowingly sell or ship a defective item. How are you modeling integrity for your employees?

Inspiration and Supply Chain Leadership

The definition of leadership includes inspiring employees to do their best every single day. How do you inspire your employees? Some leaders provide inspiration via their charisma. They have the personality that by their very presence inspires the employees.

Other leaders must work at inspiring their employees. One very effective method that has worked for me is actually doing the job. In Kuwait, my method of inspiring my distribution center personnel was to get on a forklift and actually do the work. Forklifts were a precious commodity in the early days of the Theater Distribution Center in Kuwait. In fact, they were a very valuable and rare commodity. The number of forklifts available and actually working meant that we could not afford for a forklift to sit idle. Thus the inspiration for the title of my first book, *The Forklifts Have Nothing to Do! Lessons in Supply Chain Leadership.* In a distribution center, especially one that is handling up to 290 inbound and outbound trucks daily, there is always something for the forklifts to do!

When I would see a forklift sitting idle, I would hop on the truck and start moving supplies or loading trucks. This started as a way to keep supplies moving and expedite the loading of trucks. The corollary benefit that I discovered was that it provided inspiration to the forklift operators to want to get back on the truck and show "the old man" that they could move more products or load a truck faster. In fact, we even had some friendly competitions to see who could load the trucks the fastest. No one wanted to be beat by an old colonel. What started as a way to see what the conditions were and how they affected the soldiers, and also a way to ensure that the forklifts did not sit idle, turned into a very effective inspiration tool.

At the National Training Center, the way to inspire my soldiers to do more maintenance jobs was to do some jobs myself. I had some of the best maintenance technicians in the U.S. Army in John Medlin, Joe Madrid, and Paul Barnes. I had them give me some training and check my work, and then set out to try to inspire my soldiers and the soldiers training at the National Training Center to do more work in less than ideal conditions. Imagine a ten-acre concrete pad in the middle of the summer in the Mojave Desert — not exactly a motivating factor for working outside. My goal was to get the units to complete the maintenance on several thousand pieces of equipment in the fastest time possible.

I offered the unit that set a new record for completion time the opportunity to paint its distinctive unit patch on the wall of the maintenance briefing building. This started as a way to motivate the units and turned into a great inspiration for all the soldiers in the units. Almost immediately, every unit started setting a new record for completion times. Over the course of six months, the completion times went from between twelve and fourteen days to six days; and by the end of a year, the completion times were down to five days. A simple idea that cost us absolutely nothing to implement because the units provided the paint produced a 64 percent decrease in completion times in less than a year. The bigger benefit was that soldiers were reunited with their families a week earlier. This provided a big boost in morale and inspired the soldiers to continue to want to do better.

Inspiration comes from knowing your employees and knowing what makes them tick and what motivates them. My experience is that the best way to inspire employees is do the work and apply simple, no-cost incentives to achieve the goals.

Where do you get your inspiration? Not everyone is inspired in the same way. Inspiration is tied to knowing your employees, and is discussed in Chapter 4. Every leader should aspire to provide inspiration to his/her employees every single day. Dare to inspire someone today!

Integrity and inspiration are linked because the modeling of behavior that exudes integrity, coupled with success, inspires employees to emulate those actions and act in a manner that exudes integrity while reaching new levels of excellence.

As I finished writing this chapter, a new example of leadership integrity — or lack thereof — surfaced. Several very senior U.S. Army general officers came out before the 2008 U.S. Presidential elections and stated that they did not believe that they should vote and had, in fact, not voted since they were colonels. This means that when they were, by regulation and law, appointing "Voter Assistance Officers" and supposedly encouraging their soldiers to vote, they themselves were not voting. Thus, they were practicing the old lack of integrity concept of "Do as I say, not as I do." This seeming lack of integrity in actions and words is in direct conflict with the philosophy of another senior Army officer and later U.S. President, Dwight D. Eisenhower, when he said, "The supreme quality in a leader is unquestionable integrity. Without it, no real success is possible." When you are modeling supply chain integrity and personal integrity for your employees, do not subscribe to the "do as I say" model.

Jackie Robinson once said, "A life is not important, except in the impact that it has on the lives of others." What impact does your life have on the lives of others? Are you inspiring them to be the absolute best that they can be, or are you inspiring them to cut corners, shave time off the clock, and waste company money? Are you inspiring your employees to "be all they can be?"

Supply Chain Intelligence

In the Foreword to Alan Axelrod's *Eisenhower on Leadership*, Peter Georgescu is quoted as saying, "Intelligence can be defined as the ability to observe seemingly nonexistent patterns."[13] In today's supply chains, there are nonexistent patterns of demand, patterns of customer actions, and possibly patterns of supplier support. What are the seemingly nonexistent patterns in your supply chain?

In the U.S. Army's Combat Studies Institute's book, *Studies in Battle Command*, Dr. Jerold Brown wrote about the Battle of the Little Big Horn:

> The plan seemed to guarantee success; a similar plan had worked eight years earlier when Custer attacked a village on the Washita in Oklahoma. Thus, the plan was fully consistent with known facts and the personality of the commander. ... All of the decisions Custer made on 25 June 1876 until the time he climbed Weir Peak were consistent with his vision of the battle he would fight. ... If his intelligence had been correct and if the enemy behaved, history might have written a different chapter that day. Custer's vision on this day was faulty.[14]

Is your supply chain intelligence faulty? What is supply chain intelligence? Supply chain intelligence is all of the information that is needed to make informed decisions in today's supply chain. Sun Tzu warned us to not rely on rustics. In this case he was speaking of acting on unreliable or faulty data. Custer made his decisions based on his points of reference and ignored signs that pointed to a situation different from his previous intelligence reports. This proved disastrous for his soldiers and his unit.

Military history is full of examples of leaders taking action based on flawed intelligence. In today's supply chains, acting on flawed intelligence may prove just as disastrous as Custer's actions. The difference is that Custer's actions resulted in the loss of lives, and acting on flawed supply chain intelligence results in the loss of jobs, loss of credibility, and possibly the loss of the company itself.

Is there a Supply Chain Leadership IQ? Just as there is an Intelligence Quotient, there is also a Supply Chain Leadership Intelligence Quotient. This is a measure of how much a leader knows about supply chain operations, coupled with a leader's knowledge of leadership and how to use these concepts to lead supply chain employees to new levels of commitment and excellence.

How do you develop supply chain intelligence? The concepts of self-development and employee development discussed in Chapter 10 will assist a leader in developing and applying his/her Supply Chain IQ.

Summary

Leaders must have integrity to be credible in any environment. General Dwight D. Eisenhower said, "The supreme quality in a leader is unquestionably integrity." Without integrity it is possible to get ahead, and you can fool people into thinking that you are honest and straightforward. In previous days, this type of person was known as a confidence man, or "con man" for short. Eventually your integrity will make the difference between success and failure.

Your Supply Chain IQ will have an impact on your credibility and ability to be a leader in today's supply chain operations. And, if you have a high degree of Supply Chain Leadership IQ as defined in this chapter, you will have the ability to motivate and inspire your employees to achieve new levels of excellence.

Notes

1. Leadership and Force Development, Air Force Doctrine Document 1-1, February 18, 2006, pp. 1–5. www.dtic.mil/doctrine/jel/service_pubs/afdd1_1.pdf.
2. Supply Chain Integrity, A Basis to Upset Gray Market Distribution, TransportGistics, Inc. White Paper.
3. In the mid-1980s, the U.S. Army started conducting Internal Vulnerability Audits and Assessment of supply chain operations to ensure that applicable laws and regulations were being followed and to ensure that supply activities were policing themselves and aware of vulnerabilities.

4. http://www.nascar.com/2008/news/headlines/cup/02/13/dearnhartjr.cmears. johnson.jgordon.engines.ap/index.html. Accessed February 14, 2008.
5. Faergemand, Jacob and Jespersen, Dorte. September–October 2004. *ISO Management Systems Magazine*, p. 21.
6. Examples of supply chain integrity issues can be seen in the following incidents in the past several years:
    *   Peter Pan peanut butter was recalled due to *E. coli* in 2007.
    *   *E. coli* from fresh spinach caused all fresh spinach to be taken off the shelves in 2006.
    *   Chi-Chi's restaurants never really recovered from the *E. coli* infections caused by green onions that were apparently fertilized with human waste in 2003.
    *   Taco Bell was forced to recall meat due to *E. coli* in 2005.
    *   Several meat packers recalled potentially contaminated meat in early 2008.
7. Handfield, Robert. Ethics and Supply Chain Management in a Global Environment, http://www.ncsu.edu/scrc/public/s2hottopics.html. Accessed on February 20, 2008.
8. An example of the stress on terrorism is the initiation of terrorism insurance. U.S. President George W. Bush signed the Terrorism Risk Insurance Act (TRIA) into law in November 2002 to stimulate business investment that had slowed to a trickle after the events of September 11, 2001. The law creates a three-year federal program that backs up insurance companies and guarantees that certain terrorist-related claims will be paid.
9. Aberdeen Group, Press Resources. Supply Chain Risk Still Increasing While the Market Stands Still, http://www.aberdeen.com/press/releases/press_release.asp? rid=69. Accessed February 13, 2008.
10. Katz, Norman A., CFE, Katzscan Inc., www.ncof.com/NCOF2008/custom/Katzscan SupplyChainFraudWP.pdf, p. 2.
11. There was security at every link in the supply chain to prevent attacks or sabotage of military supplies. This security and 24/7 visibility of the distribution center and staging areas prevented pilferage of supplies.
12. Gilmore, Dan. March 11, 2008. *Global Supply Chain News* weekly newsletter.
13. Axelrod, Alan. 2006. *Eisenhower on Leadership*, Jossey-Bass, a Wiley imprint, San Francisco, p. v.
14. Brown, Jerold E., Ph.D. Custer's Vision, *Studies in Battle Command*, Combat Studies Institute, Fort Leavenworth, KS 66027, pp. 75–78.

Chapter 12 Questions

1. What happens when the standards of integrity for your company or your supply chain conflict with the standards of integrity in the countries that you are outsourcing to or selling in?
2. How do you handle the violations of your supply chain integrity by employees?
3. Is the big contract worth compromising your integrity in the short run?
4. What is the cost of catching employees "DOING something right?"
5. How are you inspiring your employees today?
6. What can you do to improve your personal Supply Chain Leadership IQ?

Chapter 13

P4
Professional Pride, Planning, Passion, People

A day lived without doing something good for others is a day not worth living.
—Mother Theresa

Why are these virtues, qualities, or leadership enablers important? People form the foundation of everything we do in supply chains. Without quality people and quality leadership, an organization can never reach a level of professional pride that exudes from tier-one companies. Without a passion for providing quality goods, services, and supply chain support, and without a passion for taking care of people inside and outside our supply chains, as leaders we can never reach the top. And, without proper planning for the future, we cannot sustain our supply chains or meet customer needs.

Professional Pride

To set a baseline for the discussion of professional pride, we need to define both *professional* and *pride* and then develop a definition for *professional pride*. "Professional" is defined by Dictionary.com as "a person who is expert at his or her work."[1] The goal of every supply chain leader should be to develop subordinates who are experts at their work while developing personal professionalism.

A profession is defined as "the body of people in a learned occupation."[2] Using this definition, the business of supply chains is definitely a profession. Another definition for profession is: "the collective body of persons engaged in a calling."[3] Usually, a calling in this definition is in a field such as religion or medicine. However, it could very well be argued that supply chain leadership is a calling. Being a supply chain leader is not easy. In fact, as one former Commander of the Operations Group at the National Training Center[4] was fond of saying, "Logistics (this was before we became supply chains) is tough; if it was easy we would call it tactics." This comes from the senior operations tactics trainer in the U.S. Army. So, I would argue that to be in such a tough profession as supply chain leadership is a calling.

Having defined "profession," let's take a look at "pride." Pride is defined as "a high or inordinate opinion of one's own dignity, importance, merit, or superiority, whether as cherished in the mind or as displayed in bearing, conduct, etc.; the state or feeling of being proud."[5] Another important aspect of pride is defined as "a feeling of self-respect and personal worth."[6] Professionalism, being those traits demonstrated by a professional, is defined as "the expertness characteristic of a professional person."[7]

Armed with these definitions of profession, professionalism, and pride, we can define professional pride as being proud of what a person is doing in his/her profession that produces a feeling of the state or feeling of self-respect and personal worth. How do you get this feeling of self-respect from what you do in a supply chain? This is where leadership is critical.

Ask a forklift operator in your distribution center what he/she does for a living and the normal response is, "I drive a forklift, pick up stuff over there and take it to the shelves, and then take stuff from the shelves to the shipping area." How exciting is that? One of the advantages of being in Kuwait was that I was not constrained by the OSHA requirements for forklift certification. So, I got to jump on forklifts and drive them around. I have to admit that driving a forklift eight hours a day can become very dull and droll. However, my challenge was to find out how to make the driving of a forklift outside in the desert sun fun and exciting.

I am sure that more than a few people thought I was crazy; I could have stayed in the safety of the headquarters building. When I first gathered together my full-time crew after a couple of weeks of using temporary labor, I asked them if they understood the mission that we were chartered with at the Theater Distribution Center. Most of the responses centered on "moving stuff," "off-loading the trucks," and "loading trucks." When I explained that they were the key to getting the right stuff to their comrades who were preparing for combat and that what they were doing was critical to success of the operation, the soldiers and the civilian workers got all pumped up and felt professional pride in what they were tasked to do. Try it; the same philosophy will work in your distribution center when the workers realize that they are the final link between manufacturing all over the world and customers that keep them employed and keep the country moving forward.

Folks in the distribution center certainly thought I was crazy as I drove around the yards in a forklift singing and appearing to enjoy myself. The goal was to get them excited about what they were doing. To make the challenge work, I had to show my employees that working on the forklift could be fun. To do that, we created some competitions in unloading the inbound trucks and loading the outbound trucks.

When I was at the U.S. Army's National Training Center the second time, I worked hard to get the employees to develop a sense of professional pride. I used the example of the old Hanes™ underwear commercials when the inspector proudly proclaimed, "It doesn't say Hanes until I say it says Hanes." I also used the example of Mercedes-Benz. I noticed when I lived in Germany that all the Mercedes automobiles had a little sticker on the front windshield that had a replica of the signature of Mr. Daimler.

What I wanted to convey was that the owner of Daimler-Benz was so proud of the workmanship of the company that he was willing to sign his name on the vehicles. My goal was to get my mechanics to have the same sense of professional pride. I used the philosophy that even if they did not sign their names on the vehicle after it was repaired or rebuilt, that their name was on it as the mechanic. The mechanics had been beaten down so long by the previous supply chain executives and bosses that they were afraid that the program was designed to try to play "Gotcha!" This is what happens when you establish a climate where the supervisors constantly try to find someone doing something wrong. What I wanted was a climate where the workers were so proud of the work that they were doing that they were willing to put a little three-by-five-inch card in the window of the vehicle or in the maintenance folder that accompanied every piece of equipment for which we were responsible.

It took several months to change the attitudes of the workers. When they finally realized that the focus was customer support and a statement of pride, they "bought into" the program. We discovered a collateral benefit. With some low-density units such as logistics units, chemical units, and air defense units getting the opportunity to train at the National Training Center more often, they would look at the card in the window or folder to see if they recognized the mechanic from a previous training exercise. We started hearing, "I know this mechanic; if he (she) says that they fixed it, I know it is good to go." This type of comment helped improve the self-esteem and professional pride of the workers. The next development after the increase in professional pride was a comment such as, "I had this same vehicle the last time and I know it is a good one."

I have tried a similar system in distribution centers. In distribution centers, I like to use what I call "Name the Aisle." Many readers of this book have used the same technique. In this method for developing and improving professional pride, I recommend placing placards at the ends of the aisles with signs that look like that in Figure 13.1.

This is a simple-to-implement program that I learned from General Chuck Mahan, former Chief Supply Chain Officer for the U.S. Army. The program accomplishes several things when properly implemented. The first is that the people whose names are on the placards all of a sudden start taking pride in "their aisle," and the second is that I have found that it creates some friendly rivalries in the distribution center between the workers who own the different aisles. The workers' professional

Aisle: _____

Maintained by: _____

_____

_____

Inventory accuracy: _____

Figure 13.1 "Name the Aisle" placard.

pride pushes them to want to have the best-looking, cleanest, and straightest aisle with the highest inventory accuracy in the center. The workers go out of their way to keep their aisles straight and orderly.

Your name is going on the product or service whether you want it to or not. Therefore, it is the leader's responsibility to model a behavior that encourages putting your name on the product out of professional pride.

Planning

Leaders at all levels must have a plan for where they want to be, where they want the company to be, and how to get there. Remember the old cliché: "If you are failing to plan, then you are planning to fail." There is a lot of truth in many old clichés and definitely in this one. Your subordinates are watching you to see what your plan is and how you react to glitches in the plan. We previously discussed the After Action Review (AAR) as a method of reviewing the operation after it has been conducted. One of the steps in the AAR is "What was the plan?". If there is no plan, it does make the AAR easy but it does make operations very difficult.

The U.S. military uses an adaptation of the Scientific Decision Making Model as a planning guide. The Military Decision Making Process is a good guide to planning any operation from a military operation to releasing a new product or opening a new distribution center. We now discuss this model as a way to develop plans for any organization.

The steps in the Military Decision Making Process for developing plans include:

1. *Receipt of the guidance from the boss.* Hopefully, this guidance will be clear, concise, and detailed enough to allow for the development of a plan.
2. *Analysis of the guidance.* Exactly what is the boss asking us to do? Are there specific things that he/she requested? Are there tasks that we know from experience must be done (implied tasks) to accomplish what the boss is looking for? Do we have the proper resources and personnel on hand to accomplish these tasks?

3. *Determine alternative methods or actions to accomplish the tasks.* Here is where the actual planning process begins. Is there more than one way to accomplish what the boss wants? There is always the option of doing nothing. I have seen more than one organization that subscribes to the theory of waiting to see if guidance changes or if there are leadership changes before the plan must be developed. I do not recommend this action. What are the alternatives to doing nothing or continuing to do things the way we have always done them?

4. *Analyze the alternatives.* Do they meet the boss' intent? Are they really different? What are the cost benefits or customer benefits to each of the alternatives? What are the risks associated with each of the alternatives, and do the risks outweigh the benefits?

5. *Compare the alternatives.* This may be as simple as constructing a payoff chart or benefits chart. The payoff chart is a relatively easy table to construct. Simply list the alternative actions and the associated benefits or profits/costs to determine which of the alternative methods of meeting the boss' guidance are best for the company and the customer while accomplishing the goals of the boss.

6. *Develop the plan.* At this point it is time to start the planning based on the comparison in the previous step. Once a solid plan is developed, it is time to get the boss' approval and start the operational process based on this plan.

Leaders must model a behavior of planning and preparation for their employees to see and emulate. How often have you seen a supervisor become upset with an employee for something that should have been part of the operational plan that the supervisor or his/her boss should have developed?

Passion

What is *passion*? There are two very good definitions for passion that fit into this discussion of modeling supply chain leadership. The *Random House Dictionary* gives us: "any powerful or compelling emotion or feeling; a strong or extravagant fondness, enthusiasm, or desire for anything."[8] *The American Heritage® Dictionary of the English Language* gives us another applicable definition for passion: "boundless enthusiasm."[9] Both definitions are useful in developing the personal traits that are necessary for modeling supply chain leadership.

As exciting as supply chain operations are and as critical as this business is to the overall success of the company and the overall success of the country, how can you not be passionate about what we do for a living every single day. What other job in the world allows you the opportunity to see the results of your work every single day? How can you not be passionate about this business?

What in the world is not supply chain related? What other industry can honestly say that it is so intertwined with every citizen and every customer? A leader must have not only the passion for this business, but that passion must also be visible to everyone that he/she comes in contact with. And that passion must be real, genuine, and sincere. As stated earlier, you can fake almost anything and any feeling— but someone will surely recognize it if you are faking enthusiasm. Enthusiasm for this industry is easy to have and easy to model for your employees. How can you expect your employees to be enthusiastic when dealing with customers if all they see is negativity and just going through the motions from their leaders?

Another aspect of passion that is critical for leaders is a passion for people. Because leadership is a people business, all real leaders have a passion for taking care of their people. I once had a boss who commented that I was too concerned for my employees. I am not sure that such is possible. If we are truly in a people business— and we are—then we must have a true passion or "boundless enthusiasm" for taking care of our employees.

People

As stated repeatedly throughout this book, we are all in the people business whether or not we want to admit it. That is why we are concerned about leadership— because people need leadership.

In the people business, it is important to have interest in your employees. What motivates them? What are their interests outside the office or distribution center? Just as everything in the world is related to supply chains, everything we do is related to people. We are leading people, working for someone who is leading people, delivering products to people, receiving products from people, or manufacturing goods for people. Without people we are not in business.

What your employees and co-workers are looking for is a leader who is concerned about his/her people, is passionate about taking care of them, and treats them just as he/she would want to be treated.

People are the foundation of our supply chains.

Summary

People, *passion*, and *professional pride* are the cornerstones of everything that we do as leaders. We must model a behavior that shows that we are not only concerned about our employees, but also that we are so passionate about our profession that we have professional pride in what we do and what our employees do in meeting customer expectations — every single day.

As leaders we have to *plan* for where we want our organization and our employees to be in the future and make that plan clear for everyone to see and understand.

Notes

1. professional. Dictionary.com. *Dictionary.com Unabridged (v 1.1)*. Random House, Inc., New York. http://dictionary.reference.com/browse/professional. Accessed: October 10, 2008.
2. profession. Dictionary.com. *WordNet® 3.0*. Princeton University. http://dictionary. reference.com/browse/profession. Accessed: October 10, 2008.
3. profession. *Webster's Revised Unabridged Dictionary*. Retrieved October 10, 2008, from Dictionary.com Web site: http://dictionary.reference.com/browse/profession.
4. The Operations Group is an organization of approximately 1000 soldiers who are considered the best of their rank and profession. These soldiers serve as the senior trainers for the U.S. Army in all aspects of operations.
5. pride. Dictionary.com. *Dictionary.com Unabridged (v 1.1)*. Random House, Inc. http://dictionary.reference.com/browse/pride. Accessed: October 10, 2008.
6. pride. Dictionary.com. *WordNet® 3.0*. Princeton University. http://dictionary.reference.com/browse/pride. Accessed: October 10, 2008.
7. professionalism. Dictionary.com. *WordNet® 3.0*. Princeton University. http://dictionary.reference.com/browse/professionalism. Accessed: October 10, 2008.
8. passion. Dictionary.com. *Dictionary.com Unabridged (v 1.1)*. Random House, Inc., New York. http://dictionary.reference.com/browse/passion. Accessed: October 10, 2008.
9. passion. Dictionary.com. *The American Heritage® Dictionary of the English Language, fourth edition*. Houghton Mifflin Company, 2004. http://dictionary.reference.com/browse/passion. Accessed: October 10, 2008.

Chapter 13 Questions

1. Do we have a clear plan of where the company is going and how to best meet our customers' expectations? Do we really know what the customer wants as we develop our corporate plans?
2. What can I do to improve the professional pride of my employees? Will the "name the aisle" technique work in my distribution center?
3. Is my company really modeling that people are the foundation of our success?
4. Do we have passion for what we are doing in my company?

APPLICATIONS

Section I provided us with a foundation on leadership by providing definitions, a historical look at leadership and a way of looking at leadership using the basics of Six Sigma to guide leadership actions. Section II provided a detailed look at the attributes of leadership that are critical to supply chain and operational success and the attributes that are critical to modeling supply chain leadership. Section III provides the methodology for benchmarking leadership in supply chain operations by providing a set of supply chain leadership metrics and a scale for benchmarking. In addition, Section III provides guidance on using your position as a supply chain leader to coach, teach and mentor your employees. All leaders need to be coaches, teachers and mentors. Although coaches should be leaders and teachers, this is not always the case and the goal of this section is to provide guidance on being a coach for employees at all levels of an organization without having to go to an outside consultant or outside executive coach.

A coach's legacy lives in the lives of the players that they impact. A supply chain leader's legacy is very similar—the legacy of a supply chain leader is the lives of the employees that they impact as they model leadership and coach their employees. The application of the attributes in Section II and the metrics and benchmarks in Section III will assist in developing your employees and your legacy. A teacher's legacy is his/her students. A supply chain leader is a teacher, a coach and a mentor while modeling leadership and developing his/her legacy. A supply chain leader's legacy is his/her employees, customers and suppliers that emulate the leadership modeled by the supply chain leader using the attributes in Section II and the Applications in Section III.

Chapter 14

Coaching, Teaching, Mentoring

> Leaders create an environment that makes others successful.... not
> themselves.
> —Lou Holtz

Mentoring, coaching, teaching, and leading — relationships with subordinates, relationships with superiors — the base is relationships. How do you coach if you do not know the fundamentals? The number-one team sport is business; and in today's supply chains, coaches are critical to success.

One of the fastest growing professions in the United States is *executive coaching*. Based on my observations and discussions with executive coaches, this is really counseling and facilitating. Would you hire a head football coach for your alma mater who never played football? Then why would you hire an executive coach for your supply chain executives who has never been a supply chain executive? Do you hire an executive coach simply because he/she is certified? Certified by whom? What are the criteria for certifying a coach? As a former strength coach, I know the examination process for becoming a Certified Strength and Conditioning Specialist is intense and detailed. Becoming an APICS Certified in Production and Inventory Management is a very detailed process. Both of these certifications also require a background in the business prior to applying for certification.

Based on observations of executive coaches at professional conferences over the past several years and talking with these coaches over dinner about their work and how they got into the business, only a few of them have ever been in any business except the coaching business. At one supply chain conference, none of the coaches had ever worked in the supply chain business and yet were "bragging" about how much they made from coaching — one of these coaches boasted that he "rarely" traveled to the executives who he coached and did all of his "coaching" over the phone. I recently had lunch with a business coach from Italy who was shocked that any coach would practice over the phone.

Now, I am not against executive coaches; in fact, I believe that they do serve a viable purpose and fulfill a need in the business world. However, you must have experience to teach as a coach. As a coach of the National Championship Powerlifting team, I had to have knowledge of the sport. Having competed for 20 years helped, but not every lifter becomes a coach. In fact, very few become coaches. In baseball, not every player is qualified to become a coach — again, it goes back to studying the sport. This is not book learning that qualifies one to be a coach — leadership and executive coaching is the same — it takes more than book learning as a facilitator or counselor to become a coach of executives. A former executive or supply chain professional is better qualified to be a coach for supply chain executives.

> Coaching is the art and practice of inspiring, energizing, and facilitating the performance, learning and development of the player.
> —Myles Downey[1]

The coach, like every other leader, provides purpose, direction, and motivation to his/her team while establishing a winning attitude and working hard to make the team and the team members better as individuals and as a team. Just as one size does not fit coaching, it does not fit leadership either. Coaching is a people business, and people need one-on-one leadership and coaching to help them build self-esteem.

Where do coaches turn to for guidance? They turn to a mentor. And they must be careful who they turn to as a mentor. Make sure your mentor is qualified, experienced, and can be objective in providing your guidance. A mentor asks where do you want to be and how we can get there. A mentor may be a friend but that is not sufficient qualification to be a mentor. Your mentor must have successful experience in your field in order to provide you with guidance.

All of us need mentors but a good mentor should have experience in the business that we are in in order to assist us. Can you lead if you have not done it on the ground? You certainly would not go to a financial mentor/advisor with no experience in the financial markets; but supply chain leaders continue to go to executive coaches/mentors with no experience in the supply chain arena. You have to understand the environment to lead and build confidence; you have to understand the business to mentor supply chain leaders.

Over the past several years I have observed the operations of pit crews and crew chiefs in NASCAR and Indy Racing League races at Daytona and the Kansas Speedway. The crew chiefs serve as head coaches for the teams. As head coaches, they have to mentor the pit crews and the drivers to ensure that everyone is working together. During a long race or a tight race, the crew chief is the person who keeps the team focused and keeps the driver focused. The same is true for supply chains.

Coaches of supply chain personnel must keep the team composed and operating, regardless of the market situation and the customer demands.

Coaches must have a strong background in the area that they are coaching. Strength coaches who appear to know what they are doing have more credibility. A slim strength coach is not necessarily unable to coach strength athletes; in fact, one of the smartest strength coaches I have encountered was the coach at a college in Florida about twenty years ago. But let's be honest, a person who looks strong has more instant credibility as a strength coach than a slim coach. A strength athlete will have more confidence in the guidance of a coach who at least appears to know how to apply the techniques of the sport.

At the same time, just being a good athlete does not qualify a person as a good coach. Some of the better coaches in sports were not superstars during their playing careers. And some superstars never make it as coaches. The same is true for supply chain coaches. Every good supply chain manager will not necessarily be a good supply chain leader or supply chain coach. However, success in the supply chain world will add to the credibility of the coach.

As the owner of an all-star cheerleading gym, I have encountered more than a few cheerleading coaches. I am always amazed at cheerleading coaches who yell at their athletes for not being able to do a stunt that the coach is obviously not able to do based on his/her skills. Or coaches who are obviously out of shape and trying to tell young girls how to get in shape.

I have spent a career observing football coaches, baseball coaches, and basketball coaches. The coaches who had a great career did not always make the best coach. But the coaches who studied the game were the ones who were successful. Every coach is a teacher and mentor; however, not every teacher is a coach or a mentor. Bill Self, the head coach of the University of Kansas 2008 National Champion Basketball Team, showed during the championship year that he was indeed a coach, teacher, and mentor. Two players experienced tragic deaths in their families and Coach Self helped them get through the tragedies and continue working — not only toward a national championship, but also toward completing a degree and graduating.

There is a difference between coaching and merely teaching. What is that difference? Every coach is a teacher but not every teacher is a coach. Being a teacher, although a very honorable profession, does not automatically qualify a person to be a coach; there are additional skills, knowledge, and abilities required to pull a team together to function as a team.

Coaching

How do you coach? What is a coach? Dictionary.com gives us several applicable definitions of coach. A coach is a person who trains an athlete or a team of athletes; and to coach as "to give instruction or advice to in the capacity of a coach; instruct."[3] The tie between coaching and teaching is the word "instruct"; just as a teacher instructs, so does a coach. A supply chain coach is a person who trains a supply chain professional or team of supply chain professionals.

Not everyone can be a coach. Not everyone needs to be a coach. There are some people who feel like they do not need a coach, and some people and some athletes are not coachable; this is a sad situation. And, not everyone needs to be coached the same way. There is no one-size-fits-all coaching method.

Jimmy Valvano used to tell the story of his first coaching experience. He wanted to be just like what he had read about Vince Lombardi. What Coach Valvano quickly found out when he blew his first coaching speech was that there was only one Vince Lombardi and what others have found out in the ensuing years is that there was only one Jimmy Valvano. Does that mean that you cannot learn how to coach by watching other coaches? No, what it does mean is that what works for one coach does not necessarily work for another. And, as Coach Bobby Knight discovered, what works in one situation at one job may not work in a different situation or at a different job.

Athletic coaches must know their players. Supply chain coaches are no different. A good coach must know the strengths and weaknesses of every person on his/her team. A good supply chain coach must know the strengths and weaknesses of every member of his/her team. The coach has the tough job of putting the best team on the floor. As many coaches have found, that may not mean putting the best player or best employee on the team. It means forming a team and putting the best mix of employees on the floor to support the supply chain customer.

Every member of the team may not need the same level of coaching and teaching. This does not necessarily mean that you show favoritism to superstar employees as some managers believe. It does mean that because of their abilities, some team members do not need the same intensity of training and coaching, and some team members may need additional training and coaching. By knowing the skills of each member of the team, the coach can determine which members need to learn how to be part of a team and which members need to learn the basic skills to contribute to the team. Without this knowledge, the coach may not be successful in leading the team to new levels of excellence.

Critical coaching skills/enablers include:

- *Listening.* I know it is hard to believe but coaches do not know everything and need to listen to their team members. Coach Dean Smith of the University of North Carolina let his players tell him when they needed a break on the court. A good athletic coach depends on the honesty of his/her athletes to know when one is really hurt and needs to come out of the game or leave practice. A good supply chain leader has to listen to his/her team members to know when something is not going right or needs to be altered or changed to provide the superior level of support to the customers of the supply chain.
- *Observing.* A coach has to watch what his/her "players" are doing. Yogi Berra once said, "You can learn a lot by just watching." Yogi, as usual, was

on the mark for coaches. A good coach of a sports team or a supply chain team has to observe his/her "players" on a regular basis to know what their skills are, what their attitude is on a daily basis, and in which areas the "players" need improvement and one-on-one coaching. A coach who does not observe his/her players may not know enough about the players and the mission to ensure that the right team is "on the playing field." If you do not observe your players on a regular basis, how can you counsel them and mentor them to improve their performance? Surely you do not wait until the end of the performance period to tell the person how he/she performed? As a coach, it is imperative to observe the members of your team in order to help them set their performance goals for the year. The U.S. Army established a formal counseling program for all officers and non-commissioned officers to ensure that they were getting more than an annual performance counseling. Just having a program did not ensure that all the required counseling was performed. A later revision to the program required the counseled and the supervisor to initial and date the counseling form after each required counseling. As a coach, you owe your team members regular counseling on their performance and coaching in the areas that need improvement. You do not need a formal program like the U.S. Army has implemented, but you do need a regular session where you tell your team members what they are doing well and what areas they need to focus on and improve, based on your observations. Just as you cannot coach a basketball team from behind a computer screen, you cannot coach your supply chain team from behind the computer screen. You have to get out of the office and observe your team in action — even if that means traveling to another location to see them in action.

■ *Knowing your "players."* This coaching enabler is tied to the previous enabler of observing your players. A coach has to know his/her team members — what motivates them, what demotivates them, what their interests are outside of work. All these answers are important to coaching the members of your team to achieve the next levels of excellence. Observing your team members enables you to know the answers to these questions. With these answers in hand, the coach can develop a game plan and a performance improvement plan. The goal of the coach is to get his/her team members to perform at their optimal level. Not every team member is going to be a superstar; in fact, if all the team members are superstars, the job of the coach is extremely difficult. Look at the Los Angeles Lakers; they put together a team of the best players in the league; the New York Yankees did the same thing in baseball as the Lakers did in basketball. The results were similar. The teams of superstars did not produce the championships that the teams of stars working as a team did in their respective sports. The same is true in the supply chain arena. Your goal as a coach is not to have a team of superstars but a superstar team of members who you motivate to get the absolute best from every single day. As a coach, your success is only as good as you are today. If you do not believe that, just ask Joe Torre of the Los Angeles Dodgers. After one of the most successful coaching careers as the manager of the New York Yankees, he was let go. This was nothing new for him. He suffered the same fate as manager of the Atlanta Braves. All he did in Atlanta was return the Braves to respectability but did not win a World Series title.[4] What you did yesterday is not important to your customers; what you do for them today is all they will remember. Glory fades quickly — you have to earn your customer victories and respect every single day in the supply chain world. A customer only remembers that this time around he did not get everything he ordered on time and complete.

■ *Dealing with people.* Coaching, just like leadership, is a people job. Dealing with people — on your team and on the other teams, to include the customer team — is critical to success in the supply chain coaching job. There is not a one-size-fits-all technique for dealing with people, just as there is no one-size-fits-all employee. The successful coach has to know his/her team members and use that knowledge to deal with each one of them as individuals. Remember that a leader does not lead an organization; a leader leads the people who make up the organization. A coach does not lead or coach a team as much as he/she coaches/leads the members who make up the team. When you coach individuals, you must deal with them during the good times and the not-so-good times.

■ *Handling people.* This is related to the previous enabler of dealing with people. Coaches have no choice; they have to deal with individuals every day while shaping the team to accomplish the supply chain mission. Handling people is similar and yet different. Handling people deals with the aspect of how to release team members when they either will not or simply cannot perform to the level that is expected after receiving the proper training and coaching. This is quickly becoming a lost skill with some coaches because the mores of society seem to dictate that everyone who goes out for a sport in high school will make the team. They may not ever get in a game but they will be part of the team. In the supply chain world, it is similar; because of the shortage of skilled employees willing to work in our distribution centers, we sometimes attract those employees who are not at the top of the education world. Some of these workers will be superstars in the distribution center, and some will require additional training and coaching one-on-one to get to an acceptable level. Some employees in every walk of life and business just want a paycheck. In the supply chain world, as we continually try to cut costs, an employee who just wants a paycheck without working for it is a true liability. How you deal with such an employee requires great skill. One professional coach commented that the greatest draw on his leadership skills came when he had to tell a player that he was no longer needed on the team. Your biggest challenges as a supply chain coach will be in handling the team member who is not making the grade and in handling the problem employee. Professional sports simply trade the player who is a problem to another team. Many companies use the technique of "don't send them away mad, just send them away" by moving the problem employee to another section of the company. This is not always a bad thing. Sometimes a star employee will degrade in performance because of the lack of leadership skills of their new first-line supervisor. In these cases, knowing the employee will enable the coach to "trade" the team member to another section. In this case, moving the employee is acceptable. But moving an employee to another department just to get him/her out of your area does not solve the company's problem. There are times when it becomes necessary to move an employee out of the company. Just as a sports coach keeps statistics on his/her players, a good supply chain coach needs to keep statistics on his/her supply chain "players" so that when a promotion is warranted, he/her can support the promotion; and when a termination is necessary, the coach needs to have the proper documentation of all the coaching and training that the employee has received but did not apply in his/her performance. How do you deal with a problem employee? It is critical in today's society to document everything in writing. This serves two purposes. The first purpose — and in my opinion the most important — is to ensure that employees have the counseling in writing when they are told the areas that they are deficient in and to ensure that they understand the consequences of continuing in the same behavior or lack of performance. The second purpose is to show the review board, if necessary, why a person was terminated and the paper trail with the employee's signature showing that you have tried to improve their performance and behavior and have provided not only the counseling on a routine basis, but also the necessary training to enable the employee to improve. When it comes to dealing with problem employees, do you simply send them to additional training courses to get them out of the way for awhile or because they are expendable to your operations? Unfortunately, this happens way too often. Instead of sending the employee who really deserves the additional training that will lead to promotions in the company, we send J.S. Ragman to get him out of the picture for a while. This action is not fair to the company paying for the training, the deserving employee, or the problem employee. I have to admit that I once did just that. I had a problem employee who constantly whined and complained. He kept copious notes on every action that his supervisors, including me, did or said if it was contrary to his opinions. When given the opportunity to send this employee to a school that he was not truly qualified for, I did it anyway because it put him out of the picture for five months. The rest of the story is that when he returned, his job no longer existed and he was forced to find a job in a different department.

I use this example of what *not* to do. Send the deserving employee to promotion leading training and document the performance of the problem employee so you can terminate his employment and cover your backside against complaints.

■ *Dealing with superstars/egos/team/individual performance.* This enabler is tied to the previous enablers. A coach has to know his/her team members and observe them in order to know which players are team members, which players are superstars, which players' egos will be hurt if corrected in front of others, and which players can be counted on to put in an individual performance that will benefit the team and not just themselves. Some teams truly have superstars on the team. When a coach has the opportunity to coach a true superstar, the way the coach handles the superstar may be different — not because the superstar deserves different treatment, but because the superstar may not need as much coaching on the basic skills as other players. What the superstar of the team does need from the coach is guidance on being a team player. Another issue with superstars is ego. Coaches have to work with the superstars to ensure that their egos do not get in the way of team missions. This is a delicate issue for coaches and requires the coach to deal with the player on a one-on-one basis.

■ *Motivating.* One thing is certain: it is easy to motivate individuals when they are being shot at. Fortunately, most employees will never have to face that situation. However, although your employees are not being shot at like soldiers in war, they are being "shot at" by competitors and demanding customers. As a leader and coach, it is your responsibility to motivate your employees just as leaders in combat motivate their "employees." As a coach, it is necessary to motivate team members. To motivate your employees, you must know your employees, and this includes knowing what motivates each individual employee. Not every employee is motivated in the same way. What motivates an individual is what motivates that individual. The goal of the coach is to get the individuals to buy into the team's goals and make them their individual goals. When this happens, the coach is able to motivate the team and the individuals at the same time. This is the mark of success for a coach when measuring motivation — getting all of the team members to buy into the team goals, adopt those goals as their own, and use the team goals to motivate each member of the team. This is a little easier to do when you are coaching a sports team. When coaching the supply chain team, it is important to focus each employee by getting them to understand how the team/section/department goals tie into the employees' favorite radio station — WIIFM (or What's In It For Me?). When employees can plainly see what is in it for them, they will be motivated to accomplish the team goals.

■ *Understanding.* A coach, like a teacher and a leader, has to understand the members of his/her team. Stephen Covey speaks of the principle of "seeking first to understand and then be understood." Coaches must apply this principle every day. For your "players" to understand what you want from them, you must first understand why your "players" are there and, like the previous enabler, understand what it is that motivates that "player" and just as importantly, what de-motivates each "player." Why is this important? Some members of your supply chain team will be demotivated by being corrected in public. A large number will be motivated by praise in public. General Douglas MacArthur once said, "You should praise in public and chastise in private." This is good advice for leaders, coaches, and teachers. Support the goals of your "players" and work to get them in the limelight — but you have to be careful here. This is where understanding the employee is important. Some employees do not like being praised; they just want to do their job and contribute to the team's success.

■ *Focus on the positive.* There is an old country music song with a line that says "look on the good side and cut out watching for the bad." This is true in coaching. If you constantly belittle players on your supply chain team, they will lose self-esteem and will no longer contribute to the success of the team. Even when chastising employees, always finish on a positive note. This is nothing different from parenting; if you constantly put down your children (and in some cases our supply chain team members are like our children), they will quit trying. And because we have enough youth in our societies who have quit trying or at least quit trying positive actions, we need to make sure we do not demotivate our employees and coach them to achieve success based on their individual skills. Jimmy and Hallie Godfrey, coaches of the G-Force Jets All Star Cheerleading Teams, are good examples of focusing on the positive with young cheerleaders. Competitive cheerleading is a tough sport and mistakes happen in practices and competitions — much like some of our supply chain operations. These two coaches are very adept at focusing on what the young cheerleaders did well and then practicing those areas that need improvement. The same is true for our supply chain operations. Focus on what your employees are doing well, and add additional training and practice for those areas that need improvement. Always end constructive criticism of your employees with a positive, upbeat comment to keep from de-motivating them. To be successful at this, you must understand what drives the employee or team member and then provide a positive comment on one of those areas. Everybody does something right — find it and use it to your advantage in creating a world-class supply chain team.

■ *Discipline.* The effective use of discipline is critical to your coaching success and therefore to your supply chain success. When properly executed, discipline does not have to be harsh; it does have to be fair across the board. You cannot have one set of standards for discipline for your superstar employee and another set of standards for all other employees or team members. In fact, your discipline may be simply a look of disappointment. This works especially well with superstar employees. I once had a superstar deputy that lost focus because of some personal problems and almost failed in an important mission. When I confronted him about it, he became my teacher. Before I could say anything, he told me, "This will never happen again. I do not ever want to see that look again. I could tell that you were disappointed." In this case, this superstar deputy felt worse because he felt like he let me down than any lecture or chastising would have ever accomplished. If you are truly coaching your employees, you may very well get to the point where they do not want to let you down as their leader and coach, and will work harder to ensure that they do not let you down. This is a benchmark of true coaching. Watch the tapes of old University of North Carolina basketball games when Coach Dean Smith was the head coach; what you will see is the players giving their all to make sure that they did not let their coach down. The players at the University of California at Los Angeles (UCLA) did the same thing when they were playing for Coach John Wooden. Take a look at your coaching and leadership style; are you creating a sense of fear discipline and your supply chain team is motivated by fear, or have you created a world-class coaching atmosphere where your supply chain team is reaching world class because of their respect for you and their desire to not let you down?

■ *Coaching and leadership are a sacred trust between the leader, the organization, and the people.* As the coach, you are being watched constantly by your "players" to learn how to act, how to "play," and how to respond in tense situations. As the coach, you have to make an extra effort to ensure that you are always modeling the behavior that you want your players to emulate — do not take actions that will violate that trust between you and your "players." This is why the U.S. military has laws against "fraternization" between leaders and their soldiers. The military wants to prevent even the illusion that something may be happening that would violate the sacred trust between the leader and the led. As a leader and a coach, you must ensure that you do not do anything to violate that sacred trust with your supply chain team. You have to set the example for behavior, attitude, taking care of employees, and not allowing personalities or temptations to cause you do take actions or do something that will mar that sacred trust. Just as the military has rules against fraternization between leaders and their subordinates, you may want to adopt a formal or informal rule of a similar nature. Leaders of supply chain teams should be careful in their relationships with their subordinates. Does this mean you cannot eat with them or have an occasional drink with your supply chain team? Absolutely not! What it does mean is that you have to be careful not to take actions that will give the impression of partiality or actions that will break that sacred trust between you and those that you coach.

■ *Recognizing superior performance.* Superior performance may differ from one employee to another. But when a member of your supply chain team exhibits superior performance, recognize it publicly as soon as possible after the incident. Do not wait until several weeks after the incident — provide an immediate pat on the back. The same principle is true for discipline of problems or problem employees; do not wait for several weeks; make an on-the-spot correction to improve the performance.

■ *The best team vs. team of best players.* We have touched on this but it bears repeating. Do you want to field a team with the best players, or do you want your supply chain team to be the best team. Sometimes it is possible to be both. More often than not, you do not get both. Look at the Olympic basketball teams since they started using professional basketball players. Only once when using college players did the USA team fail to win the gold medal, and that one is still under contention more than thirty years later. Since the professional players starting playing and the best of the National Basketball Association players have made the team, the success has not been what was expected. Why? The team is composed of the best players available — every one of them superstars — but do they comprise the best team that can be fielded or a team of the best players. In the supply chain arena, it is just as important to pick the best team to move your product and information. This may mean not having a superstar on the team. The supply chain coach then becomes responsible for molding the team, teaching fundamentals (which are just as important for the supply chain team as they are for the sports team), and developing the team members into a cohesive team focused on the success of the supply chain team. The coach and the rest of the supply chain leadership have to determine their goal: is it short-term gains using a team of superstars, or long-term success by selecting, developing, and mentoring the supply chain team into the best team on the field, focused on supporting the customer?

■ *Getting your supply chain team to do its best every day.* How do you define success for your supply chain team? When every member of the team strives daily to do his/her absolute best — that is coaching success. This also leads to coaching peace of mind. When I was wrestling and struggling, my coach did not make the effort to redo my attitude about success. Before a critical tournament, I received a note from my Dad before he left for a business trip. I have carried the note every day since then. It is faded and falling apart but still helps me stay focused. The note simply said:

> Remember win, lose or draw, I am for you. There are a lot worse things than losing a match.
> —Daddy

The point of the note was: do your best and know that you did your best. That is all a coach can ask of his players. I have heard some managers say, "His best was just not good enough." A coach will look at the same situation and ask, "What can I do to help him improve and really reach his or her potential?" When I had the opportunity to watch a Special Olympics competition, I really learned the meaning of being happy with doing the best you can. If you ever get the opportunity to watch a Special Olympics competition, go and you will see for yourself the thrill of doing the absolute best you can.

▪ *The coach as a leader and as a teacher.* A coach recognizes that there is a certain potential that is unique to each individual. The coach is not there to make an individual feel bad or criticize an individual. I have seen this type of "coaching" lead players in many sports to quit even when they had the potential to play at the next level. The coach as a leader and teacher is responsible to bring that potential out in his/her employees. Always look for the potential, and develop your game plan to develop that potential in your supply chain team members. To accomplish this, the coach as a leader and teacher must develop his/her vision of teaching. This is the method that the coach will use to provide instruction and guidance. This instruction and guidance must be done with a specific end state in mind. Military operations always include the desired end state in the guidance. As we have all seen, sometimes the desired end state is not always accomplished; but as a coach, you must have a desired end state in mind for each of your players as you develop the guidance plans for them. Part of this end state and vision for coaching, teaching, and mentoring has to include the guidance to your team members that it may be harmful to compare yourself to others when gauging success. Just as each individual has inherent talents and potential, each individual is just that — an individual — and comparing themselves to someone else may not allow them to reach their true potential. This is where the coach comes in and provides mentoring, guidance, and individual goals for success. Listen with both ears and observe with both eyes; this is why you have two ears and only one mouth. Listening is an important skill for teachers, coaches, and leaders at all levels. Listen carefully to what your supply chain team members are saying — not only might you learn something from them, but you also might hear what they are really saying and this may help you understand what is driving them or temporarily demotivating them. You may also learn why their performance temporarily drops off and learn what you have to do as the coach to get them back on track. Contrary to what some coaches I have observed lately seem to think, our "players/athletes" or team members may have a good idea that the new coach may not have thought of yet. When I was a high school senior we got our third wrestling coach in three years. The first two coaches were former wrestlers — in fact, one was only one point away from making the Olympic Wrestling Team. The third coach was not a wrestler and had never even coached wrestling. However, he was wise enough to let us all know that he was not very smart about wrestling. What impressed me was when Coach Lehning told us, "I do not know that much about wrestling, but I do know about conditioning. So, I will get you in shape and depend on the seniors to teach the new guys how to wrestle." That year we had our best year as a team. The morale of this example is that sometimes coaches can indeed learn from their players, just as players can learn from their coaches. The choice as a new or experienced coach is yours!

- *No individual is more important than the team — this leads to success for the organization.* How can this be? Doesn't a superstar deserve special treatment, and doesn't a superstar make the difference between a good team and a great team? Maybe; but more often than not, the accomplishments of the team are more important than the accomplishments of the individual. Think about some of the modern-day sports teams that add a superstar to make the team better. The Los Angeles Lakers provide a good example — treating individuals as more important than the team has not produced recent championships. More than a few Major League Baseball players have made the mistake of thinking that their individual statistics were more important than the team winning. Some athletes have gone as far as using performance-enhancing drugs to make themselves look better while never being part of a championship team. Thankfully, there is no performance-enhancing drug for the supply chain team … just good training and good coaching/leadership to move them to the next level.

- *Good days and bad days.* As a leader, your best is needed every single day — you do not get a day off from setting the example and inspiring subordinates. Players are allowed to have "off days." Coaches and leaders are not allowed this luxury. You are on display every day, and your supply chain team is looking at you and depending on your leadership and coaching every single day. Your challenge is (just as your team strives not to let you down) to not let your supply chain team down. It is difficult. That is why not everyone is cut out to be a leader; that is why not every player becomes a coach; and that is why there is such a need for world-class leaders and coaches to bring about world-class behavior in your employees every single day.

- *Ego.* Just like other aspects of leadership, there is no place in coaching for your ego. It is important to remember that as the coach, the success of the organization is not your success but the organization as a whole. You cannot be successful as a coach if your team is not successful. If you are more concerned about your own career and success than the success of your "players," then you will never be a true success. Focusing on your own personal success can be detrimental to the team. As the coach, you have to instill the same attitude in your supply chain team.

- *In Axelrod's book entitled* Eisenhower on Leadership, *he states, "There is more to good leadership than telling people what to do."*[5] The same is true for good coaching. As a coach and a teacher, you have to show the supply chain team members how to do the work and model the behavior that you expect of them. A coach cannot simply tell his/her team what to do; that is why you will see coaches diagramming the big play on a whiteboard — he/she is showing the players exactly what they need to do. A good supply chain leader must do the same thing; he/she must show his/her "players" what he/she wants them to do.

■ *Team building.* If a leader is a coach, what is his/her role in team building? Do you think a good sports coach can have a successful team without working to mold the players on his/her team into a cohesive team? I overheard one manager state, "I am in charge and we will do it my way." That attitude not only does not help build a cohesive team that is capable of meeting the requirements to support the customer, but it also works counter to the concept of team building. To build a world-class team, the coach has to know his/her players and know what their strengths and weaknesses are. Once the coach has this knowledge, he/she can build a team that balances one player's strengths with another player's weaknesses while working with the players to improve their skills and therefore reduce the weaknesses and strengthen the team. For team building, this is analogous to Goldratt's Theory of Constraints. If you do not improve the skills of your weakest team member, you will never improve the overall throughput for the entire team. As a coach, you are responsible for teaching, leading, and mentoring your supply chain team.

Teaching

How do you teach and train? Dictionary.com defines a teacher as "a person who teaches or instructs."[2] Is there a difference between training and education? Absolutely! Education is tied to a classroom; this is where the teacher comes into the picture. Training is a hands-on process; this is where the coach comes into the picture.

Teaching requires knowledge in a specific field. The old adage that "those who can, do; and those who can't, teach" is no longer applicable in our schools or our supply chains. Those who can should be the ones who teach. Teaching credibility, especially in the supply chain world, comes from a combination of knowledge and the ability to transfer that knowledge to someone else. This is why teaching is more difficult than most people realize. When I started teaching graduate courses in operations management and logistics management, I automatically assumed that because I knew the subjects, that teaching them would be easy. Nothing could be farther from the truth. Just knowing the subjects and having experience in applying the subjects and techniques did not automatically make me ready to teach. I had to master the technique of transferring the knowledge to others. Transferring knowledge to others and having them demonstrate a mastery of the subjects is what makes teaching a special profession and is why many so-called experts have trouble teaching subjects that they have mastered.

Training is the process of applying the education process to real-world situations. It can be earning by doing, as is the case with learning martial arts techniques and procedures. A person can read all the books in the world on how to do martial arts forms and techniques but without the hands-on training in a martial arts school or under the tutelage of a martial arts master, it will not be a complete learning experience.

The story portrayed in the movie *The Karate Kid* is a classic example of the dichotomy between education and training. The character Daniel, played by Ralph Macchio, is shown practicing techniques he read in a karate book, but it was not until he was trained in the arts by Mr. Miyagi that he was able to master the techniques.

The same is true in supply chains. We all need formal education courses in supply chain techniques and updates as provided by professional organizations such as APICS, the Warehousing Education and Research Council (WERC), and the Council of Supply Chain Management Professionals (CSCMP) in their professional development programs and annual conferences. At the same time, supply chain professionals need hands-on training programs and refresher programs and professional mentoring/coaching to remain at the top of their games.

Mentoring

How do you mentor someone else? A mentor, according to Dictionary.com, is "a wise and trusted counselor or teacher; or, an influential senior sponsor or supporter."[6] Mentoring is a long-term commitment that does not end when a supply chain team member leaves your team. A mentor–mentored relationship is a long-term relationship that equates to coaching and teaching over the long term. Every leader is a mentor to someone, and all leaders need a mentor to guide them in their career progression. Just as you need someone to give you advice and counseling as you develop as a leader, your subordinates also need a mentor. A good mentor–mentored relationship continues long after the work relationship ends.

However, it is also possible to have a mentor that you have never actually worked for but who has worked in your particular industry. In the case of the supply chain leader, he/she would want to look in the supply chain field to find a mentor if a previous or current boss is not available. An executive coach is not your mentor — you want your mentor to be from your professional field.

When you develop a mentor–mentored relationship, you have formed a bond that will benefit the mentor by keeping him/her involved in the business and will benefit the mentored by helping him/her become a better supply chain leader. Seek to find someone who needs a mentor and volunteer to assist him/her. But remember that being a mentor is much more than simply giving advice or stating your opinion — you become a long-term advisor and confidant of the person you are mentoring.

As the mentor, you may very well find yourself in the role of advisor, teacher, and coach. It is not easy work and should not be taken lightly when someone comes to you and asks for you to mentor him/her. It is a win–win situation for all involved.

Summary

There is a difference between teaching, coaching, and mentoring. Some of the skills are contained in all three but the difference is what sets the coaches apart from other professions. I was once accused of calling a person "merely a teacher." Teachers hold a special place in my heart; my life was shaped by some of the teachers and professors under whom I studied. Both of my daughters are preparing to become teachers, and I have dedicated my life to teaching. Being a teacher is a very honorable profession and requires a great deal of patience and knowledge. However, being a teacher does not automatically qualify a person to be a coach.

The foundation for teaching/coaching/mentoring, like leadership, is people skills. Remember that whatever business you think you are in, you are in the people business. And people need teachers to show them what to do and how to do it; they need coaches to get them to perform at the next level; and they need mentors to guide the coaches and leaders to keep them continually moving to the next level.

Notes

1. Downey, Myles. 2003. *Effective Coaching: Lessons from the Coach's Coach, second edition*, Texere.
2. Dictionary.com. Accessed July 26, 2008.
3. Dictionary.com. Accessed July 26, 2008.
4. For the record, after Joe Torre was fired as the manager of the Atlanta Braves, they became one of the worst teams in the National League. Sometimes, simply winning is much more important than winning championships.
5. Axelrod, Alan. 2006. *Eisenhower on Leadership*, Jossey-Bass, a Wiley imprint, San Francisco, p. 78.
6. Dictionary.com. Accessed July 26, 2008.

Chapter 14 Questions

1. How can I apply the enablers of coaching to my profession?
2. What do I know about the members of my team that will enable me to be a better coach, teacher, and mentor to them?
3. Who do I mentor, and do I give them what they are looking for?
4. Am I modeling the correct attitude, behavior, and integrity to be a successful coach for my supply chain team?

Chapter 15

Benchmarking Leadership and Leadership Metrics

> Leaders must never make the mistake of believing that they lead a supply chain, a company, or an organization. What they lead are the people, the individuals who make up that supply chain.

Can you really measure and benchmark leadership? You can benchmark coaching by looking at the win–loss column, by looking at the number of players from a high school program who move on to college teams, or by looking at the number of college players who move on to professional teams. And you can even benchmark coaching by looking at the number of assistant coaches who move on to become head coaches. These measures show the success of the coach in training athletes or training and preparing coaches for promotion. In the U.S. Army, you could benchmark the leaders by the number of junior officers who rose through the ranks or the number of junior officers who stayed in the Army past their initial commitments. Another measure of senior leaders was how many company commanders were later selected for battalion command and how many of a brigade commander's subordinate battalion commanders were selected for future brigade commands.

Benchmarking Leadership

Using Six Sigma (as discussed in Chapter 3) as a baseline for measuring and benchmarking leadership, we can look at the attributes and values of leadership and establish a scorecard for leadership success. Just as in Six Sigma, the first step in establishing the scorecard for leadership is to define the customer. In leadership, the internal customer is the employee. The external customer is the user of your product or service. Both of these are important to the measurement of leadership.

The internal customer metrics include employee morale, retention, and employee advancement. The importance of motivating employees is one of the common threads woven throughout this book. The benchmark for employee morale is employee retention and productivity. Employee retention is a direct reflection of leadership. In the U.S. Army, every unit is measured monthly, quarterly, and annually on its retention rates. Why? Every unit in the Army has a goal for how many of their volunteer soldiers they should retain or reenlist. The quality of the leadership of the entire organization is reflected in these retention rates. The Army understands that a certain number of young soldiers come into the Army for the educational benefits and to learn a skill, and will — regardless of the leadership experience — leave the Army after their initial three-year tour. The Army also realizes that the nature of being a soldier is not for everyone.

The U.S. Army, like any large organization, also understands that the experience of the soldier and the concern shown by his/her leaders have a direct impact on the decision to stay in the Army. Unlike civilian organizations, a simple two-week notice does not work in the Army. Therefore, the units with quality leaders experience high retention rates. The same is true in commercial supply chain organizations. Companies that routinely motivate employees and companies that routinely exhibit a true concern for their employees have high levels of employee retention. The Toyota North American Parts Distribution Center in Ontario, California, is a good example of this. The last time I visited that center, they had been open for more than ten years and had a retention rate of over 96 percent.

That is the benchmark of retention. Conversely, I worked with another distribution company in Southern California that had a 50 percent retention rate. Why? Although the leadership and management preached employee satisfaction, the pay scale was almost half the average in the area for trained forklift operators. Guess who was paying for the training for all of their competitors. As soon as their operators received their OSHA certifications, they left for one of the competitors.

What was the difference between Toyota and the other distribution center? The leadership at Toyota sought to understand the employees, what the employees needed to meet basic needs, and what was needed to keep them happy at the plant. Harley-Davidson in Kansas City is another good example of keeping employees happy. Their employee retention rates are very close to the Toyota rates. In fact, at last count, the Kansas City Harley Davidson plant had somewhere around 10,000 applicants for every posted opening. That is another good metric of leadership and employee retention — the number of people who want to join your organization.

As you develop your metrics with your internal and external customers in mind, the question that should drive your process is: What do the customers expect of me?

Leadership Metrics

Supply chain leadership metrics include:

- *Employee morale.* What is the morale of your employees? This is usually easy to ascertain as you walk around the organization. General Matthew B. Ridgeway knew this when he assumed command of the U.S. Forces in Korea. His staff told him that the morale of the soldiers was good. Knowing that morale would not be good if he had been pushed back from the Yalu River to the 38th Parallel after months of intense combat in the Korean winter, General Ridgeway walked the lines and talked to the soldiers. No surprise; what he found out was that morale was at an all-time low. In today's economy with companies outsourcing supply chain functions at a record pace, when was the last time you walked the line and talked to the employees about their morale? In 1995, I had the unenviable task of determining which of my employees were "key and essential" as the U.S. Government basically went out of business for a couple of days without a budget or a continuing resolution to cover all government functions. I had to look each of my employees in the eye and tell them if they would be coming to work the next day. Some of my fellow directors chose the easy way out and declared all their employees "key and essential" to their operations. Needless to say, employee morale was at an all-time low at this point when the majority of my employees were told not to come to work until they heard otherwise. Luckily, the Congressional stalemate only lasted a couple of days and everyone was able to come back to work. The moral of this story is much like the moral of the Ridgeway story. You have to talk to the employees — in good times and bad times — to determine the real morale.
- *Employee productivity.* Are you getting more productivity from the same workforce, or is their productivity declining? If it is declining, have you done a cause-and-effect study to determine why it is declining? Is it something that you can or did impact as a leader? When measuring productivity, it is important to look at how you are measuring your employee productivity. As in all areas of leadership, there is not a one-size-fits-all measurement. My heavy equipment mechanics usually did not produce at the same levels as my light equipment mechanics because of the complexity of the heavy equipment. You have to know your processes and your employees to effectively set productivity measurements. Regardless of how you measure the productivity, if the productivity declines during your watch as a leader, it may be an indicator of your leadership.
- *Employee retention rates.* How many of your quality employees are you retaining? When I was in the U.S. Army, one of the metrics that we were measured against at every level was retention rates. Like the

U.S. Army, you may not need to retain every single employee. Some employees need to go for various reasons. However, the ability to attract and retain a quality workforce is a key metric of the leadership of the organization. Employees want to stay in an organization that has quality leaders and one that takes care of and is concerned about its employees. A company, division, or section that continues to have low retention rates is a good indication of a leadership problem. It could be that a good manager was promoted into a leadership position without any leadership training. Why are the employees leaving the organization? Once a metric is in place to measure employee retention rates, an analysis of why can be conducted.

■ *The "climate of command."* In the U.S. Army, quality units have an outside organization conduct a "climate of command" survey that asks the soldiers a series of questions about how they are treated, their perceptions of how discipline is handled in the unit, and how often they see the unit's leadership in the offices, shops, warehouses. This same concept is applicable to your organization. In your company, this would more than likely be called a "climate of leadership" or "climate of management" survey. The principles are the same: to find out the problems and perceptions of your organization. If the leaders of your organization have walked the processes, visited the shops and sections of the distribution centers or offices, then there should be no surprises when the results of the surveys are compiled. One company had an assistant manager with three department supervisors. Of the three supervisors, one stayed, one left, and one departed on indefinite sick leave for mental stress. Not exactly a good "climate of command." This is why companies need leaders to lead the managers. In fact, this particular assistant manager was overheard saying, "You don't understand; I am not paid to do any work. As an assistant manager, I am paid to delegate." This attitude was part of the reason for the poor work climate in the organization and why this assistant manager will always be a manager and not a leader. Another part of the "climate of command" is based on caring for your employees. In the introduction to *The Carolina Way,* the great University of North Carolina Coach Dean Smith stated, "The most important thing in leadership is truly caring."[1] Another example of not creating a good "command climate" was passed on to me recently by a colleague. This individual had been asked to speak at a professional development conference. The individual's boss consented to the presentation but then, two weeks prior to the conference, called the individual in and said, "I know this is good for you and good for the organization, but it is not good for me. Therefore I am not going to let you go to the conference." Now, because there was nothing really important going on other than "holding the boss' hand," what kind of climate does this type of leadership convey to the workers? "Good leaders create a work environment that is like a family, where people care for one another, help one another, celebrate the success of their fellow workers.... If you want to motivate people to work hard, work together and work smart, help them to be successful. People give back what they receive."[2] Does this describe your organization? If not, why? Another example of setting the right climate is punctuality. How often have you worked for a boss who made you wait for an appointment? How often have you waited for one or two people to show up to start the meeting? What signal does that send? Basically, you are saying that my time is more important than yours. I had a boss who was always thirty minutes behind schedule and kept everyone waiting for appointments and meetings. If the meeting starts at 8:00 a.m., everyone should be seated and ready for the meeting to start. This includes the leader of the meeting. In fact, my position was always to be there before the meeting in enough time to speak to everyone at the meeting before it started to make them feel part of the meeting and make them feel important. One way to improve the climate at work is to make the work fun. Every employee should look forward to coming to work everyday. Establish a climate where employees are excited to come to work everyday. Do this and watch your company's productivity improve.

■ *Promotion rates.* What are the internal promotion rates? Of course, this metric does not work if your policy is to only promote from within your organization. Are your employees being trained for responsibility at the next level? Just promoting the employee is not a measure of success if the employee is not trained for the responsibilities of the next level. How many of your first-line supervisors and mid-level leaders come from within your organization? How many of your mid-level leaders move on to another company? Why? Once you promote an individual from within, do you give him/her any additional training to ensure success at the next level? The hardest promotion of all is going from "one of the guys on the line" to being the supervisor. If your policy is to promote from within, I recommend that the person promoted be moved to a new section after completing training geared to ensure his/her success at the next level.

■ *On-time delivery.* This is a customer satisfaction metric but is indicative of your leadership. If you are measuring on-time delivery from the customer perspective, then as a leader you should be concerned about making sure that your organization is meeting customer expectations. If you are modeling the correct behavior for your employees to emulate, then your employees will be concerned about meeting customer expectations and the on-time delivery metric will reflect your leadership and influence on the employees.

■ *Perfect order fulfillment: the quality of your products/perfect order fulfillment rates.* Wait a minute, aren't these measures of quality, not leadership? Yes and no. If you are leading your employees, if they understand the company standards and how meeting these standards impacts them, then these quality/customer service measures are also measures of your leadership. Employees with quality leadership strive to produce quality products, ensure that the distribution center is clean, meet customer delivery standards, and ensure that the order is complete. Companies that have delivery or order fulfillment problems can usually trace these problems to poor leadership or a lack of leadership interest/involvement.

- *Employee courtesy.* How do you treat your employees? Why do I ask? Because the way you treat your employees will manifest itself in the way your employees treat each other and how they treat your customers. Some 2000 years ago, the "Golden Rule" was established — "Do unto others as you would have them do unto you."[3] Regardless of your religious background or beliefs, this principle of leadership is a good standard to use. As long as you treat your employees with courtesy, it will show in their customer service.

- *Employee burnout.* I know this one may be a bit harder to measure but it is a key indicator of leadership. All too often, leadership is responsible for this burnout. I have heard leaders preach about the need to have a balance between work and family but never practice it themselves. One such senior leader tried to set the example by leaving the office every day at 6:00 p.m., but from the times on the e-mails the next day it was quite obvious that he was not spending time with his family. Eventually, the workers started staying later in the evening and coming in earlier in the mornings to work the e-mails and projects that the boss created at all hours of the night. The result was employee burnout and a loss of real productivity. What behavior was this leader modeling? It was not what he was preaching about balance between work and family life. Taking the work home and doing it at the house is not spending quality time with your family, and your employees will notice what you are really doing and emulate that behavior.

- *Distribution center cleanliness.* The appearance of your distribution center/operation is a metric of your leadership. Habitually, distribution centers or other operations with quality leaders are easy to spot when you walk through them. How? Watch the supervisors, distribution center managers, and workers. When they see something out of place, they pick it up and put it where it should be. This could be an item on the wrong shelf or trash on the floor. Next time you walk through your operations, watch how the employees respond to trash or items out of place.

- *Customer retention rates.* This is also known in some companies as "customer churn." How can customer retention be a leadership metric? Very simple. Sam Walton said that the attitude you show your employees will manifest itself in the attitude your employees show your customers in two weeks or less. The attitude that you model for your employees and the concern that you show for your employees and your customers will affect the rate at which you attract and retain customers. Just as employees want to work for a company that respects them and is concerned about them, customers want to feel appreciated. Customers that feel appreciated (this could be as simple as a thank-you e-mail for their business, as Joseph A. Banks did for me on a recent suit purchase) will want to continue doing business as long as the product is considered a quality one and the customer service shows concern from the entire company.

■ *Employee recognition.* Another form of recognition that is inherent in Six Sigma leadership is recognizing employees for doing a good job. In coaching, it is important for coaches to tell their athletes when they are doing a good job in the sport or are practicing/playing well. How often do you praise your employees? Not empty praise or false praise, but honest and sincere praise for what they are doing. As you measure your leadership, what is the ratio of praise to criticism?

■ *Dedicated training.* Training is usually a hands-on approach to learning or reinforcing learning and classroom education. How much time do you set aside each week, each month, or each year for employee training? Maintaining a fully trained and competent workforce is the responsibility of leaders at all levels. There is no one-size-fits-all standard for how many hours you should have. This is driven by the quality of the workforce, the experience of the workforce, and the employee retention rates. Obviously, the lower your employee retention rate, the more training you will have to have to bring the new employees up to standard. Once you get your workforce up to standard, you cannot be content to rest on your laurels. As a leader, you must continually train your workforce to keep them up to standard. Another key factor in training is to make sure that your trainers are training the employees in the right techniques and not the shortcuts that they have learned along the way. If your trainers teach the shortcuts, the shortcuts get perpetuated and eventually the full process will be lost. And what gets trained is not necessarily the right way. One final aspect of training that is critical to measuring and benchmarking leadership is the effectiveness of your training. I mention this because I have seen organizations that have certain mandatory training requirements. Many times, these mandatory subjects are given in a manner that is not effective. If you are going to take time out of your production or distribution operations to train your employees, make sure that the training is effective.

■ *Employee pride.* Are your employees proud of the product they make or the service they provide? We discussed this topic in detail in Chapter 13. This metric is a little harder to quantify than some of the other leadership metrics. However, your employees' pride will be a direct reflection of your leadership style. Are your employees willing to put their names on their products?

- *Routines.* I once had a boss who emphasized that he wanted us to "do the routine things routinely." His rationale was that we should be able to do the routine functions of our jobs with little effort — not haphazardly or ineffectively, but in a routine manner. Further guidance and clarification from this boss was that we should be able to accomplish our routine functions without having to do additional planning or jumping through hoops to get the job done. How effective is your organization at doing the routine things routinely? Is everything always a priority for your organization? What are your normal routine functions? When I competed in powerlifting, I used the same warm-up routine for every practice, workout, and competition. Why? When the time came to put it all on the line in competitions, I did not want any surprises in getting ready. The same is true for your organization. If you and your supply chain do the same routine functions the same way every day on every shift, your performance will reflect the preparation and will be noticeable to your customers and your competition.
- *Leading from the front.* Do your employees know you? Leadership recognition is a metric of leadership. Do your employees know who you are? I know this sounds strange but here is a true story about one "leader" who I observed a few years ago. As this "leader" was showing me around his operations, we came across a couple of soldiers who were doing some work in an unsafe manner. In the U.S. Army, the entire chain of command has their pictures on every unit's bulletin board and all soldiers are supposed to know their chain of command all the way from their first-line supervisor through to the President of the United States. When we came across these soldiers, the response of my escort was, "Do you know who I am?" — as if this made a difference in why they were doing something unsafe. One soldier responded, "I've seen your picture somewhere." It was all I could do to keep from laughing. The moral of this story is to get out of the office, talk to your employees, and let them see you so that you are more than just a "picture on the wall."

Scoring the Leadership Benchmarks

How do you score this chart? Rate yourself first on a scale of 1 to 10 on each of the criteria in the scorecard. A perfect score of 160 is probably not realistic. Because there has not been a perfect person on this Earth in over 2000 years, do not expect to have a perfect score. Even Jesus had a less than perfect disciple retention score and, based on Judas Iscariot, obviously an employee morale problem. A score of 120 or higher shows that you are being an effective leader — but still have room for improvement. A score of 100 to 120 means you may need some additional leadership training, either through personal self-development or through a formal training program. A score of less than 100 means you should look at some formal leadership training to improve your skills. This is not necessarily a bad thing. Research from the Supply Chain Research Institute reveals that in most cases, the move from manager to leader in supply chain organizations is not accompanied by a formal leadership development program. All too often, organizations assume that a good manager will be a good leader, or that everyone can be a leader without being given additional leadership training.

After your personal assessment, do the same scoring for your organization as a whole. The same scoring criterion applies. If you really want to improve your organization, the next step in the benchmarking process is to have your employees score your leaders on the same scale. This will give you a 360-degree assessment of your leadership and will allow you to do a gap analysis of the leadership programs between your perceptions and the perceptions of your employees.

Armed with the results of your gap analysis, it is possible to weed out simple perceptions from facts. Although perceptions are important when it comes to your employees because perception is their reality, the facts will form the basis of your formal leadership training programs and the perceptions will assist in forming your informal programs.

Notes

1. Smith, Dean and Gerald D. Bell. 2004. *The Carolina Way,* Penguin Books, New York, p. 3.
2. Smith, Dean and Gerald D. Bell. 2004. *The Carolina Way,* Penguin Books, New York, pp. 70–73.
3. Matthew 7:12. *New American Standard Bible.*

Chapter 16

Conclusions and Final Thoughts

> You must be the change you want to see in the world.
> —Gandhi

In the supply chain world as a supply chain leader, you must model the behavior, values, and attitudes that you want to see in your supply chain. The goal of this book is to provide you with a workbook and guide to evaluating your own personal supply chain leadership as well as a framework for developing not only yourself, but also your subordinates. Using the supply chain leadership attributes, values, and enablers will enable you to lead your supply chain to new levels of excellence.

As a leader, your employees are constantly watching your every action to see how to behave, how to act, how to motivate other employees, how to treat other employees as well as customers, and how to respond to certain situations. The model that you provide will be used as the way to act and treat others, or will be used as the example of how not to act. The choice of which example you are is truly up to you.

As a leader, coach, and teacher to those who are on your supply chain team, you are responsible for their development as well as your own personal development. Use this guidebook on modeling supply chain leadership to establish world-class operations throughout your supply chain. Provide the purpose, direction, and motivation to your team and watch it constantly reach new levels of excellence. Strive to motivate someone every single day and help your employees be all that they can be and live up to their true potential. The added result or benefit of this modeling is that you will prevent motivational dysfunction in your organization, and your enthusiasm and passion will be transferred to a new generation of employees.

> There are 86,400 seconds in a day. It's up to you to decide what to do with them. —Coach Jimmy Valvano Coach of the 1983 NCAA Men's National Basketball Championship Team from North Carolina State University

You can use those 86,400 seconds to create world-class operations in your organization, or you can choose to do nothing. You can be like the dead fish and continue moving downstream and give the illusion of progress, or you can apply the values and ideas of this book and create progress and improvement in your operations. The choice is yours!

> Leadership is not taught. It is instead modeled. Real leadership is not about who gets the credit. It is about who empowers others to lead. —Dr. Emily Taylor

Model the correct attributes of leadership and empower your subordinates to lead your supply chain to the new levels of excellence. Do not be the dead fish!

Let me know your success stories from applying these ideas.

Appendix 1

The After Action Review

> The Army's After Action Review (AAR) is arguably one of the most successful organizational learning methods yet devised. Yet, most every corporate effort to graft this truly innovative practices into their culture has failed because, again and again, people reduce the living practice of AAR's to a sterile technique.
> —Peter Senge

We first mentioned the After Action Review in Chapter 2 on the discussions of leadership lessons from Sun Tzu. This Appendix will provide more guidance on preparing for and conducting an After Action Review of your operations.

Modern supply chains are inherently complex. To win in today's supply chain environments requires that we all understand how to review and evaluate our supply chain performance. The After Action Review provides us with a tool and methodology for evaluating the performance of our supply chains and the complex processes within the supply chain operations. The After Action Review (AAR) will assist supply chain leaders in identifying deficiencies, identifying strengths that need to be sustained and assist in focusing efforts on areas that are part of the core competencies of the organization. Competent supply chain leaders should understand these principles and techniques for conducting an effective After Action Review.

The United States Army Training Circular and Guide for After Action Reviews (TC 25, The After Action Review) provides the following guidance on the AAR:

DEFINITION AND PURPOSE OF AFTER-ACTION REVIEWS

An after-action review (AAR) is a professional discussion of an event, focused on performance standards, that enables soldiers to discover for themselves what happened, why it happened, and how to sustain strengths and improve on weaknesses. It is a tool leaders and units can use to get maximum benefit from every mission or task. It provides-

- Candid insights into specific soldier, leader, and unit strengths and weaknesses from various perspectives.
- Feedback and insight critical to battle-focused training.
- Details often lacking in evaluation reports alone.

Figure A1   The definition and purpose of the after action review.

The purpose of the AAR is to enable the leader's assessment of the operations. The AAR allows the workers to provide input. No matter how much time a leader spends observing the operations of his or her supply chain, there is no way that he or she can see as much of what is going on as the workers within the supply chain operations.

The goal is to compare the actual output of the operations or outcome of the operations to the intended outcome or outputs. This is accomplished by using specific observations from the workers involved in the preparation and activities. A corollary benefit of allowing the workers to provide input is that it allows them to build confidence in themselves while learning from their fellow workers. The key is to focus on what went wrong and how to prevent it from happening again. A skilled leader will not allow the workers to verbally attack each other or belittle their fellow workers or critique the work of their fellow workers. The leader has to focus the discussions to the activities that went right and wrong according to the plan.

After Action Reviews

- ◆ Are Conducted as close to the actual event or activity completion as possible or even during the activity if possible
- ◆ Focus on the intended outcomes
- ◆ Involve all participants in the discussions
- ◆ Should use open-ended questions to initiate discussions.
- ◆ Should be related to a specific standard for outputs or outcomes.
- ◆ Assist in determining organization's strengths and weaknesses

Figure A2   Goals of after action reviews.

The amount and level of detail leaders need during the planning and preparation process depends on the type of AAR they will conduct and on available resources. The AAR process has four steps:

- Step 1. Planning
- Step 2. Preparing
- Step 3. Conducting
- Step 4. Following up (using AAR results)

Figure A3 AAR steps.

Planning

Planning the AAR requires identification of what areas will be the focus of the AAR and who from those operations will be involved in the data collection as well as who should attend the AAR. Another important question that should be answered in this process is where the AAR should be conducted. Experience shows that conducting the AAR in the boss's office is not a good idea or a good way to stimulate the necessary free flow of information.

Preparation

The facilitator for the AAR needs to review the standards that the operations or activities are benchmarked against for success or failure. In addition to ensuring that observations are being made, the facilitator needs to get out of the office and make some personal observations of the activities to understand the processes and better facilitate the discussions. The last stage of the preparation step is to organize the observations and compare the observations to the benchmarks.

Included in the preparation step is a thorough review of all policies, procedures, and published guidelines for the activities that will be the subject of the AAR. This allows the observers and facilitators to be current in the operations and know what the standards are for the activity or process.

Conduct

The key to the successful conduct of an AAR is to ensure and encourage maximum participation in the discussions. The most difficult task for the facilitator is to keep the discussions focused on the actual activities and the planned outcomes, while keeping opinions out of the discussions. In order to transition to the follow-up step, a detailed set of notes must be kept of the discussion points. Keep in mind the goal of the AAR is to sustain the areas that were successful while fixing a problem and not fixing blame.

The AAR is a problem-solving process. The purpose of the discussion is to identify strengths and weaknesses of the processes, the workers, and the leaders. If statistics are available on the operations/processes, they should be used in the discussions. The discussions must focus on current activities and what was supposed to happen. It is critical to the success of the AAR to make sure that it is not a negative discussion and does not overlook any activities/processes that need to be sustained for future success.

Follow-up

Fix the problem and assign someone the responsibility of ensuring that the fix is working and does not create a bigger problem somewhere else in the organization. The follow-up step is much like a continuous process improvement program step—fix the problem then walk the process again to make sure the fix worked and did not create a new problem or constraint.

The real benefit of the AAR comes from the follow-up by taking the results and applying them to future activities or processes. When a leader or an organization implements the regular use of the After Action Review to improve and sustain operations, the leaders will improve, the employees will improve and so will the organizations.

Index

M

MacArthur, General Douglas, 34
Management, leadership, distinguished, 3–10
McNamara, Carter, 41
Measurement, methodology of, 27
Measures of effectiveness, 15
Mentoring, 181
Mergers, 15–17
Metrics, 15, 86
Metrics of leadership, 185–190
Military Decision Making Process for
   developing plans, 160–161
Modeling, 101
Modeling Leadership, 1
Morale of employees, 185
Mother Theresa, 157
Motivation, 7–8, 174
Motivational dysfunctions, 6

N

Name aisle placard, 160
NASCAR, teamwork, 168–169
National Association of Purchasing
   Management Code of Ethics. *See*
   Institute for Supply Management,
   Code of Ethics
Native Americans, 112
Nazi Party, 38
Negotiations, 121
   in professional development, 121
New Jersey, Principles of Ethical Conduct,
   45–47

O

Observation, importance of, 170–171
On-time delivery, 187
O'Neil, John "Buck," 61
Order fulfillment, 187–188
Organizational design, 146–148
Organizational foundations, 1

P

Passion, 157–163
Patton, General George, 17, 34
Penick, Harvey Penick, 70
People skills, professional development, 124
Perfect order fulfillment calculation example,
   117

Performance, superior, recognizing, 176–177
Performance evaluation, 18–20
Personal integrity, 135–137
Personal reliability, 117
Personal responsibility, 115–116
Planning, 157–163
Plato, 43
Positive, focusing on, 175
The Power of Positive Thinking, 68
Pride, of employees, 189
Principles of Ethical Conduct, New Jersey,
   45–47
Problem employee, dealing with, professional
   development, 123–124
Process map example, 91
Procurement/acquisition, ethics in, 52–57
Productivity of employees, 185
Professional development, 119–125
   areas of, 121–124
   certification programs, 120
   employee professional development,
      124–125
   hands-on practice, 122–123
   negotiations, 121
   people skills, 124
   problem employee, dealing with, 123–124
   public speaking, 122
   self-development, 119–121
   technical skills, 124
   thinking methodology, 122
   written communication, 22
Professional pride, 157–163
   defined, 157
Promotion rates, 187
Public speaking, professional development,
   122
Purchased components, routine sampling, 143
Purchasing, integrity in, 141–144

R

Raw materials and purchased components
   routine sampling, 143
   sampling, 143
Recognition of employees, 189
Reliability, 109–118
   personal, 117
Respect, 109–118
   defined, 110
Responsibility, 109–118
   personal, 115–116
Retention of customers, loyalty and, 38–40